Toxicology of Chemical Respiratory Hypersensitivity

Toxicology of Chemical Respiratory Hypersensitivity

Edited by

IAN KIMBER and **REBECCA J. DEARMAN**

Zeneca Central Toxicology Laboratory, Alderley Park, Macclesfield, UK

UK   Taylor & Francis Ltd, 1 Gunpowder Square, London EC4A 3DE
USA Taylor & Francis Inc., 1900 Frost Road, Suite 101, Bristol, PA 19007

Copyright © Taylor & Francis Ltd 1997

All rights reserved. No part of this publication may be reproduced, stored in a retrieval system, or transmitted, in any form or by any means, electronic, electrostatic, magnetic tape, mechanical, photocopying, recording or otherwise, without the prior permission of the copyright owner.

British Library Cataloguing in Publication Data
A catalogue record for this book is available from the British Library.

ISBN 0-7484-0354-X

Library of Congress Cataloguing Publication data are available

Cover design by Jim Wilkie
Typeset in Times 10/12pt by Santype International Limited, Netherhampton Road, Salisbury, Wiltshire.
Printed in Great Britain by T. J. Press (Padstow) Ltd, Cornwall

Contents

List of Contributors		vii
1	**Chemical Respiratory Hypersensitivity: an Introduction** Ian Kimber and Rebecca J. Dearman	1
2	**Allergic Asthma: an Interaction between Inflammation and Repair** Anthony P. Sampson and Stephen T. Holgate	7
3	**Occupational Respiratory Allergy** Jonathon A. Bernstein, David I. Bernstein and I. Leonard Bernstein	29
4	**Occupational Respiratory Allergy: Risk Assessment and Risk Management** Katherine M. Venables and Anthony J. Newman Taylor	61
5	**Immunobiology of Chemical Respiratory Sensitization** Ian Kimber and Rebecca J. Dearman	73
6	**Predictive Assessment of Respiratory Sensitizing Potential in Guinea Pigs** Katherine Sarlo and Harry L. Ritz	107
7	**Predictive Assessment of Respiratory Sensitizing Potential of Chemicals in Mice** Ian Kimber and Rebecca J. Dearman	121
8	**Predictive Assessment of Respiratory Sensitizing Potential of Proteins in Mice** Michael K. Robinson and Thomas T. Kawabata	135

Contents

9 Chemical Respiratory Allergy: a Regulatory Viewpoint   151
Peter Evans

Index   167

List of Contributors

IAN KIMBER AND REBECCA J. DEARMAN
Zeneca Central Toxicology Laboratory, Alderley Park, Macclesfield, Cheshire, SK10 4TJ, UK.

ANTHONY P. SAMPSON AND STEPHEN T. HOLGATE
University Medicine, Southampton General Hospital, Southampton, SO16 6YD, UK.

JONATHAN A. BERNSTEIN, DAVID I. BERNSTEIN AND I. LEONARD BERNSTEIN
University of Cincinnati Medical Center, Department of Medicine, Division of Immunology, 231 Bethesda Avenue ML #563, Cincinnati, Ohio 45267, USA.

KATHERINE M. VENABLES AND ANTHONY J. NEWMAN TAYLOR
Department of Occupational and Environmental Medicine, Imperial College School of Medicine at the National Heart and Lung Institute, Manresa Road, London, SW3 6LR, UK.

KATHERINE SARLO AND HARRY L. RITZ
The Procter & Gamble Company, PO Box 538707, Cincinnati, Ohio 45253-8707, USA.

MICHAEL K. ROBINSON AND THOMAS T. KAWABATA
The Procter & Gamble Company, Procter & Gamble Pharmaceuticals, PO Box 538707, Cincinnati, Ohio 45253-8707, USA.

PETER EVANS
Health and Safety Executive, Magdalen House, Bootle, Merseyside, L20 3QZ, UK.

Chemical Respiratory Hypersensitivity: an Introduction

IAN KIMBER and REBECCA J. DEARMAN

Zeneca Central Toxicology Laboratory, Alderley Park, Macclesfield

Work-related respiratory disease is an important occupational health problem. The most frequent diagnoses are of acute pulmonary oedema and other manifestations of inhalation accidents, allergic alveolitis, asthma, pneumoconiosis and mesothelioma. In this book attention will focus upon asthma and other symptoms of pulmonary hypersensitivity resulting from allergic sensitization of the respiratory tract. In 1989 a scheme for the Surveillance of Work Related and Occupational Respiratory Disease (SWORD) was initiated in the UK, the results of which have been reported periodically (Meredith *et al.*, 1991; Meredith and McDonald, 1994; Sallie *et al.*, 1994; Ross *et al.*, 1995). In the first of these reports, describing patterns of diagnosis during 1989, it was concluded that the true incidence of acute occupational respiratory disease in the UK was some three times greater than had been appreciated previously (Meredith *et al.*, 1991). In that year the most frequent new diagnosis was of asthma which represented 26.4 per cent of all cases (Meredith *et al.*, 1991). Some five years later asthma remained the most common form of occupational respiratory disease (Ross *et al.*, 1995). The data indicate that the agents with which occupational asthma is most frequently associated are isocyanates, flour and grain dusts, proteins from laboratory animals, colophony, wood dusts, platinum salts, glutaraldehyde, acid anhydrides and proteolytic enzymes (Meredith *et al.*, 1991; Ross *et al.*, 1995). In the period since the surveillance scheme was introduced isocyanates have remained the most important cause of occupational asthma in the UK, although as a proportion of all new diagnoses isocyanate asthma has declined slightly.

It is apparent that both chemicals and proteins are able to cause work-related asthma. While the focus of this book is primarily upon respiratory hypersensitivity induced by chemicals, it is recognized that there exists a need to identify those proteins that are able to cause sensitization of the respiratory tract. This issue is addressed in Chapter 8.

In addition to those mentioned above, other chemicals have been implicated in the induction of occupational asthma. Some of these are listed in Table 1.1 and more comprehensive details are available elsewhere (Chan-Yeung and Malo, 1993). It is important to emphasize, however, that the inclusion of a chemical in such lists does not necessarily imply that respiratory symptoms result in all instances from allergic sensitization. For many chemicals associations have been found between respiratory effects and the presence of homocytotropic (IgE or IgG4) antibodies. Such associations are not universal and there is some uncertainty about the nature of immune effector mechanisms, and indeed the requirement for immune processes *per se*, in some forms of chemical-induced asthma. The subject of this book is allergic sensitization

Table 1.1 Some chemicals implicated as human respiratory allergens

Toluene diisocyanate	(Butcher et al., 1976; O'Brien et al., 1979; Banks et al., 1986)
Diphenylmethane diisocyanate	(Tansar et al., 1973; Zeiss et al., 1980; Zammit-Tabona et al., 1983)
Hexamethylene diisocyanate	(Keskinen et al., 1988; Cartier et al., 1989; Vandenplas et al., 1993)
Phthalic anhydride	(Kern, 1939; Maccia et al., 1976; Nielsen et al., 1988)
Tetrachlorophthalic anhydride	(Howe et al., 1983; Venables et al., 1987; Liss et al., 1993)
Trimellitic anhydride	(Zeiss et al., 1977; Bernstein et al., 1982; Topping et al., 1986)
Hexahydrophthalic anhydride	(Moller et al., 1985; Nielsen et al., 1994)
Maleic anhydride	(Topping et al., 1986; Durham et al., 1987)
Reactive dyes	(Docker et al., 1987; Topping et al., 1989; Park and Hong, 1991)
Carmine	(Durham et al., 1987; Quirce et al., 1994)
Glutaraldehyde	(Burge, 1989; Jachuck et al., 1989; Ross et al., 1995)
Plicatic acid	(Chan-Yeung et al., 1973; Cartier et al., 1986; Frew et al., 1993)
Platinum salts	(Pepys et al., 1972; Murdoch et al., 1986; Brooks et al., 1990)
Chloramine-T	(Bourne et al., 1979; Dijkman et al., 1981; Wass et al., 1989)

of the respiratory tract and the immunological basis for this is considered in detail in Chapter 5.

Fewer chemicals are known to cause respiratory hypersensitivity than will induce allergic contact dermatitis. Nevertheless, respiratory allergy poses important occupational health and toxicological problems (Kimber and Wilks, 1995). A diagnosis of occupational asthma can have profound consequences for the individual. Workers are advised usually to cease exposure; continued exposure to the inducing agent after first diagnosis being associated with a significantly worse outcome (Newman Taylor, 1988; Ross et al., 1995). This in turn results in an increased likelihood of periods of unemployment, difficulty in finding alternative employment and financial hardship (Cannon et al., 1994). In a recent survey it was found that some 2 to 5 years after first diagnosis under half the patients analyzed remained with the same employer, 16% were in new employment, 23 per cent were unemployed and 13 per cent had retired (Ross et al., 1995).

Clearly there is a requirement for the accurate identification of those chemicals (and proteins) that have the potential to induce sensitization of the respiratory tract and occupational asthma and for an understanding of the conditions of exposure that will represent a risk for humans. The difficulty is that there are available currently no thoroughly validated or widely accepted predictive test methods. Important progress has, however, been made and such is described in Chapters 6, 7 and 8. The need to define risk factors for respiratory sensitization and to elucidate the variables that determine individual susceptibility and the capacity of chemicals to induce and elicit allergic respiratory hypersensitivity make this an exciting and challenging area of toxicology. It is the purpose of the present book to provide a holistic view of chemical respiratory allergy from a toxicological perspective and to this end there are contributions contained within from respiratory physicians, occupational health specialists and practising toxicologists.

References

BANKS, D. E., BUTCHER, B. T. & SALVAGGIO, J. E. (1986) Isocyanate-induced respiratory disease. *Annals of Allergy*, **57**, 389–396.

BERNSTEIN, D. I., PATTERSON, R. & ZEISS, C. R. (1982) Clinical and immunological evaluation of trimellitic anhydride- and phthalic anhydride-exposed workers using a questionnaire and comparative analysis of enzyme-linked immunosorbent and radioimmunoassay studies. *Journal of Allergy and Clinical Immunology*, **69**, 311–318.

BOURNE, M. S., FLINDT, M. L. H. & WALKER, J. M. (1979) Asthma due to industrial use of chloramine. *British Medical Journal*, **2**, 10–12.

BROOKS, S. M., BAKER, D. B., GANN, P. H., JARABEK, A. M., HERTZBERG, V., GALLAGHER, J., BIAGINI, R. E. & BERNSTEIN, I. L. (1990) Cold air challenge and platinum skin reactivity in platinum refinery workers. *Chest*, **97**, 1041–1047.

BURGE, P. S. (1989) Occupational risks of glutaraldehyde. *British Medical Journal*, **299**, 1451.
BUTCHER, B. T., SALVAGGIO, J. E., WEILL, H. & ZISKIND, M. M. (1976) Toluene diisocyanate (TDI) pulmonary disease: immunologic and inhalation challenge studies. *Journal of Allergy and Clinical Immunology*, **58**, 89–100.
CANNON, J., CULLINAN, P. & NEWMAN TAYLOR, A. J. (1994) Consequences of a diagnosis of occupational asthma: a controlled study. *Thorax*, **49**, 390P.
CARTIER, A., CHAN, H., MALO, J-L., PINEAU, L., TSE, K. S. & CHAN-YEUNG, M. (1986) Occupational asthma caused by eastern white cedar (*Thuja occidentalis*) with demonstration that plicatic acid is present in this wood dust and is the causal agent. *Journal of Allergy and Clinical Immunology*, **77**, 639–645.
CARTIER, A., GRAMMER, L. C., MALO, J-L., LAGIER, F., GHEZZO, H., HARRIS, K. & PATTERSON, R. (1989) Specific serum antibodies against isocyanates. Association with occupational asthma. *Journal of Allergy and Clinical Immunology*, **84**, 507–514.
CHAN-YEUNG, M. & MALO, J-L. (1993) Compendium 1. Table of the major inducers of occupational asthma, in BERNSTEIN, I. L., CHAN-YEUNG, M., MALO, J-L. and BERNSTEIN, D. I. (Eds) *Asthma in the Workplace*, pp. 595–623. New York: Marcel Dekker.
CHAN-YEUNG, M., BARTON, G., MACLEAN, L. & GRZYBOWSKI, S. (1973) Occupational asthma and rhinitis due to western red cedar (*Thuja plicata*). *American Review of Respiratory Disease*, **108**, 1094–1102.
DIJKMAN, J. H., VOOREN, P. H. & KRAMPS, J. A. (1981) Occupational asthma due to inhalation of chloramine T. I. Clinical observations and inhalation-provocation studies. *International Archives of Allergy and Applied Immunology*, **64**, 422–427.
DOCKER, A., WATTIE, J. M., TOPPING, M. D., LUCZYNSKA, C. M., NEWMAN TAYLOR, A. J., PICKERING, C. A. C., THOMAS, P. & GOMPERTZ, D. (1987) Clinical and immunological investigations of respiratory disease in workers using reactive dyes. *British Journal of Industrial Medicine*, **44**, 534–541.
DURHAM, S. R., GRANEEK, B. J., HAWKINS, R. & NEWMAN TAYLOR, A. J. (1987) The temporal relationship between increases in airway responsiveness to histamine and late asthmatic responses induced by occupational agents. *Journal of Allergy and Clinical Immunology*, **79**, 398–406.
FREW, A., CHAN, H., DRYDEN, P., SALARI, H., LAM, S. & CHAN-YEUNG, M. (1993) Immunologic studies of the mechanisms of occupational asthma caused by western red cedar. *Journal of Allergy and Clinical Immunology*, **92**, 466–478.
HOWE, W., VENABLES, K. M., TOPPING, M. D., DALLY, M. B., HAWKINS, R., LAW, S. J. & NEWMAN TAYLOR, A. J. (1983) Tetrachlorophthalic anhydride asthma: evidence for specific IgE antibody. *Journal of Allergy and Clinical Immunology*, **71**, 5–11.
JACHUK, S. J., BOUND, C. L., STEAL, J. & BLAIN, P. G. (1989) Occupational hazard in hospital staff exposed to 2 percent glutaraldehyde in an endoscopy unit. *Journal of the Society of Occupational Medicine*, **39**, 69–71.
KERN, R. A. (1939) Asthma and allergic rhinitis due to sensitization to phthalic anhydride: report of a case. *Journal of Allergy*, **10**, 164–165.
KESKINEN, H., TUPASELA, O., TIIKKAINEN, U. & NORDMAN, H. (1988) Experience of specific IgE in asthma. *Clinical Allergy*, **18**, 597–604.

KIMBER, I. & WILKS, M. F. (1995) Chemical respiratory allergy. Toxicological and occupational health issues, *Human and Experimental Toxicology*, **14**, 735–736.

LISS, G. M., BERNSTEIN, D., GENESOVE, L., ROOS, J. D. & LIM, J. (1993) Assessment of risk factors for IgE-mediated sensitization to tetrachlorophthalic anhydride. *Journal of Allergy and Clinical Immunology*, **92**, 237–247.

MACCIA, C. A., BERNSTEIN, I. L., EMMETT, E. A. & BROOKS, S. M. (1976) *In vitro* demonstration of specific IgE in phthalic anhydride hypersensitivity. *American Review of Respiratory Disease*, **113**, 701–704.

MEREDITH, S. K. & McDONALD, J. C. (1994) Work-related disease in the United Kingdom, 1989–1992: report on the SWORD project. *Occupational Medicine*, **44**, 183–189.

MEREDITH, S. K., TAYLOR, V. M. & McDONALD, J. C. (1991) Occupational respiratory disease in the United Kingdom 1989: a report to the British Thoracic Society and Society of Occupational Medicine by the SWORD project group. *British Journal of Industrial Medicine*, **48**, 292–298.

MOLLER, D. R., GALLAGHER, J. S., BERNSTEIN, D. I., WILCOX, T. G., BURROUGHS, H. E. & BERNSTEIN, I. L. (1985) Detection of IgE-mediated respiratory sensitization in workers exposed to hexahydrophthalic anhydride. *Journal of Allergy and Clinical Immunology*, **75**, 663–672.

MURDOCH, R. D., PEPYS, J. & HUGHES, E. G. (1986) IgE antibody responses to platinum group metals: a large scale refinery survey. *British Journal of Industrial Medicine*, **43**, 37–43.

NEWMAN TAYLOR, A. J. (1988) Occupational asthma. *Postgraduate Medical Journal*, **64**, 505–510.

NIELSEN, J., WELINDER, H., SCHUTZ, A. & SKERFVING, S. (1988) Specific serum antibodies against phthalic anhydride in occupationally exposed subjects. *Journal of Allergy and Clinical Immunology*, **82**, 126–133.

NIELSEN, J., WELINDER, H., OTTOSON, H., BENSRYD, I., VENGE, P. & SKERFVING, S. (1994) Nasal challenge shows pathogenetic relevance of specific IgE serum antibodies for nasal symptoms caused by hexahydrophthalic anhydride. *Clinical and Experimental Allergy*, **24**, 440–449.

O'BRIEN, I. M., HARRIES, M. G., BURGE, P. S. & PEPYS, J. (1979) Toluene di-isocyanate-induced asthma. I. Reactions to TDI, MDI, HDI and histamine. *Clinical Allergy*, **9**, 1–6.

PARK, H. S. & HONG, C-S. (1991) The significance of specific IgG and IgG4 antibodies to a reactive dye in exposed workers, *Clinical and Experimental Allergy*. **21**, 357–362.

PEPYS, J., PICKERING, C. A. C. & HUGHES, E. G. (1972) Asthma due to inhaled chemical agents: complex salts of platinum. *Clinical Allergy*, **2**, 391–396.

QUIRCE, S., CUEVAS, M., OLAGUIBEL, J. M. & TABAR, A. I. (1994) Occupational asthma and immunologic responses induced by inhaled carmine among employees at a factory making natural dyes. *Journal of Allergy and Clinical Immunology*, **93**, 44–52.

ROSS, D. J., SALLIE, B. A. & McDONALD, J. C. (1995) SWORD '94: surveillance of work-related and occupational respiratory disease in the UK. *Occupational Medicine*, **45**, 175–178.

SALLIE, B. A., ROSS, D. J., MEREDITH, S. K. & McDONALD, J. C. (1994) SWORD '93: surveillance of work-related and occupational respiratory disease in the UK. *Occupational Medicine*, **44**, 177–182.

TANSAR, A. R., BOURKE, M. P. & BLANDFORD, A. G. (1973) Isocyanate asthma: respiratory symptoms caused by diphenylmethane diisocyanate. *Thorax*, **28**, 596–600.

TOPPING, M. D., VENABLES, K. M., LUCZYNSKA, C. M., HOWE, W. & NEWMAN TAYLOR, A. J. (1986) Specificity of the human IgE response to inhaled acid anhydrides. *Journal of Allergy and Clinical Immunology*, **77**, 834–842.

TOPPING, M. D., FORSTER, H. W., IDE, C. W., KENNEDY, F. M., LEACH, A. M. & SORKIN, S. (1989) Respiratory allergy and specific immunoglobulin E and immunoglobulin G antibodies to reactive dyes used in the wool industry. *Journal of Occupational Medicine*, **31**, 857–862.

VANDENPLAS, O., CARTIER, A., LESAGE, J., CLOUTIER, Y., PERREAULT, G., GRAMMER, L. C., SHAUGHNESSY, M. A. & MALO, J.-L. (1993) Prepolymers of hexamethylene diisocyanate as a cause of occupational asthma. *Journal of Allergy and Clinical Immunology*, **91**, 850–861.

VENABLES, K. M., TOPPING, M. D., NUNN, A. J., HOWE, W. & NEWMAN TAYLOR, A. J. (1987) Immunologic and functional consequences of chemical (tetrachlorophthalic anhydride)-induced asthma after four years of avoidance of exposure. *Journal of Allergy and Clinical Immunology*, **80**, 212–218.

WASS, U., BELIN, L. & ERIKSSON, N. E. (1989) Immunological specificity of chloramine-T-induced IgE antibodies in serum from a sensitized worker. *Clinical Allergy*, **19**, 463–471.

ZAMMIT-TABONA, M., SHERKIN, M., KIJEK, K., CHAN, H. & CHAN-YEUNG, M. (1983) Asthma caused by diphenylmethane diisocyanate in foundry workers. Clinical, bronchial provocation and immunologic studies. *American Review of Respiratory Disease*, **128**, 226–230.

ZEISS, C. R., KANELLAKES, T. M., BELLONE, J. D., LEVITZ, D., PRUZANSKY, J. J. & PATTERSON, R. (1980) Immunoglobulin E-mediated asthma and hypersensitivity pneumonitis with precipitating anti-hapten antibodies due to diphenylmethane diisocyanate (MDI) exposure. *Journal of Allergy and Clinical Immunology*, **65**, 346–352.

ZEISS, C. R., PATTERSON, R., PRUZANSKY, J. J., MILLER, M., ROSENBERG, M. & LEVITZ, D. (1977) Trimellitic anhydride-induced airway syndromes: clinical and immunologic studies. *Journal of Allergy and Clinical Immunology*, **60**, 96–103.

2

Allergic Asthma: an Interaction between Inflammation and Repair

ANTHONY P. SAMPSON and STEPHEN T. HOLGATE

University Medicine, Southampton General Hospital

2.1 The History of Asthma

2.1.1 Asthma: Humoral or Muscular?

A respiratory illness typified by wheezing and panting first enters recorded history in Chinese medical writings from the Zhou dynasty in the first millennium BC, and these writings may reflect the knowledge and medical practice of even earlier epochs, traditionally stretching back 4500 years (Simons, 1994). A Chinese asthma remedy dating from the same period was the eating of the berries of the *Ephedra* plant, containing the undoubtedly effective bronchodilator and decongestant compound ephedrine, which still appears in modern pharmacopoeias.

Little of the exotic philosophical basis of Chinese medicine filtered through to the West, but knowledge of *Ephedra* spread via the Middle East to the cultures of ancient Greece and Rome; Pliny the Elder (d. 79 AD) prescribed its use, along with some other oral treatments, such as millipedes crushed in honey, which have since fallen out of favour. Around 400 BC, the Greek physician Hippocrates from the island of Cos suggested that asthma resulted from an excessive flow of phlegm, one of the four 'humours', into the lungs, while the Roman author Galen (2nd century AD), who had dissected extensively, would later stress the importance of the respiratory muscles. This divergence in the view of asthma as a 'humoral' (inflammatory) or 'muscular' (bronchoconstrictor) disorder has been a common theme over succeeding centuries.

Hippocratic teachings were developed further in the first and second centuries AD by the Roman writer Celsus and by the Alexandrian Greek physician Aretaeus. Aretaeus was the first to outline with tolerable accuracy the constellation of symptoms comprising the asthma syndrome, and to describe both exercise-induced wheeze and nocturnal asthma. Arab scholars transmitted ancient medical knowledge surprisingly intact to medieval Europe, but criticism of Galen and Celsus and their beliefs began to arise only after the Renaissance encouraged original thought and experimentation based on clinical observations. Sir John Floyer in his *Treatise of Asthma* of 1698 emphasized the chronic nature of asthma underlying acute attacks of wheeze, and John Swett in 1852 hypothesized that it was chronic lung inflammation which made asthmatic patients susceptible to a range of triggers. The definitive work at this time was by Henry Hyde Salter (1868), who summarized the current state of knowledge and outlined treatments, some of which, such as strong coffee, may have been effective due to the bronchodilator action of caffeine, while others including opium and cigarette smoke have long been abandoned. Salter also laid the foundations for the view of asthma as a neurotic disorder which bedevils understanding of the disease among the lay public to this day.

His contemporary Armand Trousseau, however, described the association of asthma with allergic and immune disorders such as eczema, hay fever and rheumatism. This observation, and the idea that the immune system can cause as well as prevent disease, are the true antecedents of the modern concept of asthma as a chronic inflammatory disorder associated with allergy and dysfunction of the immune system. Fundamental advances in allergy research in the twentieth century include the elucidation of the role of histamine in anaphylaxis by Sir Henry Dale, its localization to the mast cell granule by Riley and West in the 1950s and the identification of the 'reagin' of Prausnitz and Kustner as immunoglobulin E by the Ishizakas in 1967 (Simons, 1994). Together with the discovery of other mediator families, including the prostanoids and leukotrienes, these advances prompted a useful view of acute asthma as an IgE-mediated mast cell degranulation process, with the release or synthesis of mediators causing short-lived bronchospasm. This view persisted well into the late 1970s, and a reappraisal of the chronic underlying pathophysiology has only gained ground in the last 10 years.

2.1.2 Definition of Asthma

Accurate definition of a problem is often the first step towards a meaningful solution. Salter in the 1860s felt able to define asthma as a 'paroxysmal dyspnoea of a peculiar character generally periodic with healthy respiration between attacks', and surprisingly similar definitions revolving around acute exacerbations persisted until the 1960s. This rather limiting emphasis on bronchoconstriction as the central pathological feature of asthma, although useful as a basis for the development of β_2-bronchodilators and of 'mast cell stabilizing drugs' such as sodium cromoglycate, was a product of the ready availability of experimental techniques to measure lung function. The success of these techniques led to non-specific bronchial hyperresponsiveness, their most important finding, being included in the definition of asthma accepted by the

American Thoracic Society (1962). The simplistic link between hyperresponsiveness and clinical asthma has since been progressively undermined (Smith and McFadden, 1995), and by 1971 a study group set up to address the problem felt that there was insufficient evidence even to attempt a definition (Porter and Birch, 1971).

2.1.3 Asthma as an Inflammatory Disease

A crucial precursor in understanding the causes, rather than the effects, of episodic bronchospasm in asthma came in the early 1960s when pathological changes in the bronchial mucosa of post-mortem asthmatic lungs were systematically investigated (Dunnill, 1960). These changes included mucus plugging, epithelial damage, goblet cell hyperplasia, basement membrane thickening and eosinophil infiltration of the bronchial wall. The latter reawakened interest in the earlier appreciation of the links between asthma and eosinophilia in the blood and sputum.

The means to examine pathological changes in asthmatics in life largely awaited the development of fibre-optic bronchoscopy, which obviated the need for general anaesthesia and carried greatly reduced risks compared with rigid bronchoscopy. Mediators could now be measured in bronchoalveolar lavage (BAL) fluid, while the types of inflammatory cells lining the airways could be quantified and their function studied *in vitro*. In addition, examination of small bronchial mucosal biopsies provided a window directly into the structure of the dysfunctional asthmatic airway wall. These studies have proved crucial in demonstrating that the airways even of mild asthmatics are chronically

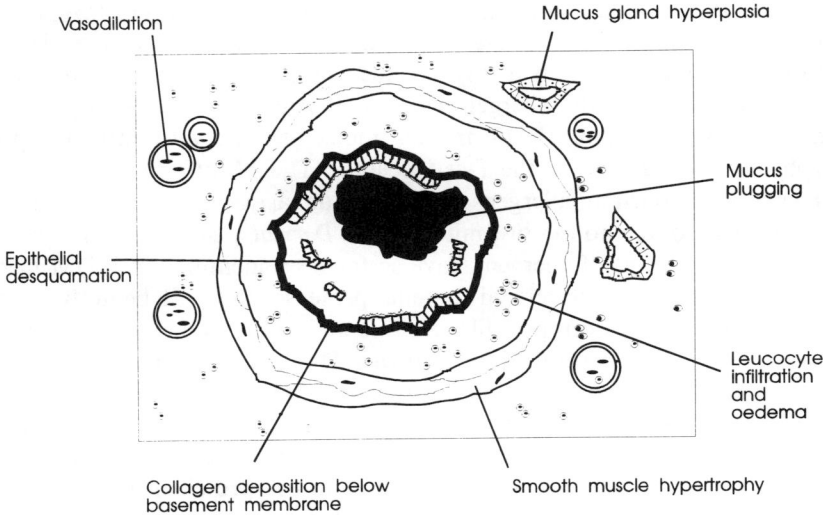

Figure 2.1 Pathological changes in the asthmatic airway.

inflamed, with infiltration of activated eosinophils and T-lymphocytes, degranulation of mast cells, epithelial damage, deposition of collagen below the basement membrane and goblet cell hyperplasia (Beasley et al., 1989; Jeffery et al., 1989; Djukanović et al., 1990) (Figure 2.1). The concept of asthma as an inflammatory disease with a progressive element has already led to a major shift in emphasis in the treatment of asthma, from relief bronchodilation with β_2-agonists to anti-inflammatory therapy with corticosteroids, and new classes of drugs that block specific families of inflammatory mediators are now coming of age.

2.2 Allergen Sensitization

2.2.1 *Atopy and Allergens*

The role of grass pollen in provoking acute episodes of asthma and hay fever was described anecdotally by Pliny the Elder in the first century AD, and more systematically by Charles Blackley in his treatises on 'Catarrhus Aestivus' (Blackley, 1873). The majority of asthma, especially in children and young adults, is linked to an inherited IgE-dependent hypersensitivity to airborne allergens (atopy), and atopy is the most important risk factor yet identified for the development of asthma. The mode of inheritance of atopy (and of asthma) is controversial. Some researchers have suggested autosomal dominant inheritance, with the genetic abnormality mapping to chromosome 11 (11q13) and that it is linked to the locus encoding the β-chain of the high affinity IgE receptor (Fc$_\varepsilon$RI) (Cookson, 1993; Sandford et al., 1993). Although highly appealing, this link has not been confirmed by other workers (Marsh and Meyers, 1992) and most authors emphasize the likelihood of multiple gene factors in the predisposition to overproduce IgE. A genetic predisposition to produce IgE to a specific allergen such as ryegrass is linked with particular phenotypes of HLA Class II DR and DP molecules, and with particular polymorphisms of the α-chain of the T-cell receptor (TCR)(Moffatt et al., 1994).

The most important allergens in provoking asthma are proteins in the faecal particles of the house-dust mite (HDM) *Dermatophagoides pteronyssinus*. At least seven groups of allergens have so far been identified in HDM faeces, with *Der p* I being a gut-derived cysteine protease, *Der p* II being lysozyme, *Der p* III being a chymotrypsin-like enzyme and *Der p* IV being an α-amylase (Stewart, 1994). That several of the known HDM allergens are proteases may provide a clue to their ability to penetrate epithelial surfaces and cause sensitization. Conversely, many other common allergens are not proteases, and allergens in general represent a bewildering variety with few common features other than their foreignness (e.g. plant and insect proteins), molecular size (most are proteins of MM 12–50 kDa), and the ability to breach physical defence mechanisms of the allergic subject. The latter combines a number of factors, including the allergen's stability and water solubility, its prevalence in the environment and its packaging in particles which allow suspension in the air and access to the mucous membranes of the eyes and nose and the bronchial epithelium. Apart from HDM, common aeroallergens include components in wind-dispersed pollens from grasses (rye, couch, timothy), weeds (ragweed, mugwort) and trees (birch, alder), in mould spores (*Aspergillus*), in bird feathers and in animal danders and urine (cat, dog, rodents). Ryegrass pollen contains several biochemically-diverse allergenic proteins, which together may comprise around 30 per cent of the mass of the pollen grain, indicating that the packaging of an allergen is more important for sensitization than its biochemical activity (Holgate and Church, 1993).

Respiratory problems may also be provoked in individuals sensitized to substances encountered in the workplace. Occupational allergens include animal and plant proteins associated with farming, animal breeding, brewing and baking, and bacterial enzymes used in detergents. Other occupational allergens are relatively small molecules which may sensitize only after haptenization. These include metals and their salts (platinum, vanadium, chromium), various other industrial chemicals (isocyanates, chloramine T, PVC) and pharmaceutical products such as antibiotics (penicillin, cephalosporin) and, ironically, the bronchodilator drug salbutamol. Only some individuals in the workplace are usually affected, again indicating that the genetic predisposition of the host is an important and little understood factor.

2.2.2 Environmental Factors Affecting Allergen Sensitization

Exposure to HDM allergen within the first year of life is a prime determinant of the prevalence of asthma and bronchial hyperresponsiveness at the age of 11 years in children with atopic parents, and the degree of exposure to Der p 1 correlates with the age of onset of wheeze in childhood (Sporik *et al.*, 1990). Allergen avoidance in the home has been reported greatly to reduce the risk of wheeze and atopic dermatitis in infants of atopic mothers (Arshad *et al.*, 1993). The risk of early exposure to allergen may extend back into gestation, as T cells from cord blood of neonates who subsequently develop asthma or atopic dermatitis show enhanced proliferative responses to egg and milk proteins and to aeroallergens (Warner *et al.*, 1994). Gestational allergen exposure may be affected by adjuvant factors including maternal smoking, which raises cord blood IgE levels and impairs infant lung function (Royal College of Physicians, 1992), and by foetal nutrition. Undernutrition in the later stages of gestation has been hypothesized to be a particular risk factor for foetuses on a fast growth trajectory, leading to disproportionate growth of the head and possible effects on thymic development resulting in IgE hyperresponsiveness (Godfrey *et al.*, 1994).

In adulthood, respiratory viral infections are one of the commonest causes of acute asthma exacerbations. Viral infection may damage bronchial epithelium allowing inhaled allergens access to the bronchial mucosa. Infection with respiratory syncytial virus (RSV) appears to predispose infants to IgE hyperresponsiveness and asthma (Russi et al., 1993). Exposure of bronchial epithelium to air pollutants including nitrogen dioxide, sulphur dioxide, ozone and elements in cigarette smoke may synergize with viral infection to promote allergen sensitization. Much further research is required to elucidate the interaction of environmental and genetic factors in the causation of asthma and to throw light on the reasons for the rapid and dramatic increase in asthma prevalence and mortality in affluent countries.

2.3 Inhaled Allergen Challenge as a Model of Asthma

2.3.1 Mast Cells and the Early Asthmatic Response

Challenge of sensitized subjects with inhaled allergen has been a vital experimental approach in asthma research since the first bronchoprovocation studies were performed by Max Samter in Berlin in 1933. Allergen inhalation results in an early bronchoconstrictor ('asthmatic') response (EAR) at 5 to 10 minutes lasting up to an hour and, in about half of subjects, a late bronchoconstrictor response (LAR) starting at 2 to 3 hours and lasting for 12 to 24 hours (Pepys, 1973).

The critical interaction of allergen is the cross-linking of specific IgE molecules bound to high affinity IgE receptors ($Fc_\varepsilon RI$), expressed on mast cells, basophils, dendritic cells and probably eosinophils, and to low affinity receptors ($Fc_\varepsilon RII$; CD23) expressed on macrophages, eosinophils, and platelets (Holgate and Church, 1993). Cross-linkage leads to the release of a range of inflammatory mediators, both preformed (histamine and tryptase) and newly synthesized (the cysteinyl-leukotriene LTC_4 and prostaglandin D_2). Histamine is a bronchoconstrictor and vasodilator acting at H_1 receptors on bronchial and vascular smooth muscle. Tryptase, which comprises 20 per cent of the total protein of the mast cell granule, is a 130 kDa tetramer which increases microvascular permeability, upregulates adhesion molecules, activates eosinophils and promotes proliferation of fibroblasts and epithelium (Walls et al., 1995). Prostaglandin D_2 is a bronchoconstrictor and mucus secretagogue acting at the thromboxane receptor (TP). The cysteinyl-leukotrienes LTC_4, LTD_4 and LTE_4 comprise the slow-reacting substance of anaphylaxis (SRS-A) and act at specific receptors to produce long-lived bronchoconstriction, microvascular leakage and mucus secretion (Arm and Lee, 1993). They may also induce bronchial hyperresponsiveness and act as eosinophil chemoattractants (Hay et al., 1995).

Clinical trials with specific antagonists, and measurements of mediators and their metabolites in BAL fluid, blood and urine, have demonstrated that the bronchoconstrictor element of the EAR is due overwhelmingly (approximately 80 per cent) to cysteinyl-leukotrienes, with minor components contributed by histamine and possibly prostaglandins (Holgate et al., 1996), reinforcing the view that the EAR is a mast cell-dependent response resulting from cross-linkage of receptor-bound IgE on the surface of primed mast cells.

2.3.2 The Late Asthmatic Response as a Model for Clinical Asthma

Cysteinyl-leukotrienes and histamine also play a major role in the LAR, as judged by trials of antagonists and synthesis inhibitors (Holgate et al., 1996). Unlike the EAR, β_2-bronchodilators only partially reverse the LAR, showing that the airway narrowing is only partly due to airway smooth muscle contraction, with the remainder possibly a result of oedema and mucus plugging of the airway. The LAR also displays two features crucially associated with chronic or clinical asthma, namely an increase in non-specific bronchial responsiveness (Cockcroft and Murdock, 1987) and an influx of inflammatory leucocytes, particularly eosinophils (persistent) and neutrophils (short-lived) (Metzger et al., 1987). The LAR has thus come to be regarded as a model of the inflammation underlying chronic asthma, and studies of the mechanisms responsible for the eosinophil influx account for much of the recent history of asthma research.

A transient decrease in eosinophils in the circulation 2 hours before the LAR has been described (Cookson et al., 1989), and this is followed within 24 hours by an increase in circulating eosinophils, a greater proportion of which have the activated hypodense phenotype (Calhoun et al., 1991) and express surface activation markers. At the same time, increased eosinophil numbers are observed in BAL fluid (De Monchy et al., 1985) and in bronchial mucosal biopsies (Frew et al., 1995). These results suggest that following allergen challenge eosinophils are selectively recruited into the lung from the vasculature, and replenished by an increased proliferation or release of eosinophils in the bone marrow.

Eosinophil recruitment, activation in the vasculature and proliferation, may all be caused by the action of eosinophilopoietic cytokines including granulocyte/macrophage colony-stimulating factor (GM-CSF), interleukin (IL-) 5 and IL-3 formed in the lung and acting as paracrine or even endocrine factors. GM-CSF and IL-5 immunoreactivity have been found in the plasma and BAL fluid of patients with acute severe asthma (Brown et al., 1991; Sur et al., 1995). Alternatively, these factors may only prime circulating eosinophils to respond more robustly to non-specific chemoattractants such as leukotriene B_4 and IL-8 (Sehmi et al., 1992; Shute, 1994). The latter hypothesis is supported by the finding of a significant neutrophil influx alongside eosinophils in the early stages (5 to 6 hours) of the LAR (Montefort et al., 1994a). Chronic eosinophilia in the asthmatic lung may depend more on the ability of eosinophilopoietic cytokines to reduce eosinophil apoptosis than on any specific process of recruitment. Indeed, IL-5 is a potent eosinophil survival enhancing factor and IL-5 transcripts are clearly identifiable in bronchial biopsies in the later stages (24 hours) of the LAR (Bentley *et al.*, 1993; Frew *et al.*, 1995) and in chronic asthma (Robinson *et al.*, 1992).

2.3.3 Adhesion Molecules

Vascular endothelium is the initial barrier to leucocyte recruitment in the lung and our understanding of the cell adhesion molecules (CAMs) by which leucocytes negotiate this barrier has advanced rapidly in recent years (Figure 2.2). Carbohydrate ligands (e.g. sialyl Lewis X) on leucocytes initially interact loosely with lectin-binding regions of P-, L- and E-selectin, causing the leucocyte to roll along the endothelial wall. Selectin expression on endothelium is modulated after allergen challenge. Autacoids including histamine, leukotrienes B_4 and C_4, and platelet-activating factor (PAF) transiently upregulate P-selectin on endothelial cells within an hour of exposure, while a slower expression of E-selectin occurs in response to the primary inflammatory cytokines IL-1, tumour necrosis factor (TNF-α) and interferon-γ (IFN-γ) (Gundel *et al.*, 1991; Vonderheide and Springer, 1992). Leucocyte rolling is arrested by interaction of leucocyte integrins including leucocyte function antigen 1 (LFA-1) (CD11a/CD18), found on neutrophils, eosinophils and lymphocytes, with

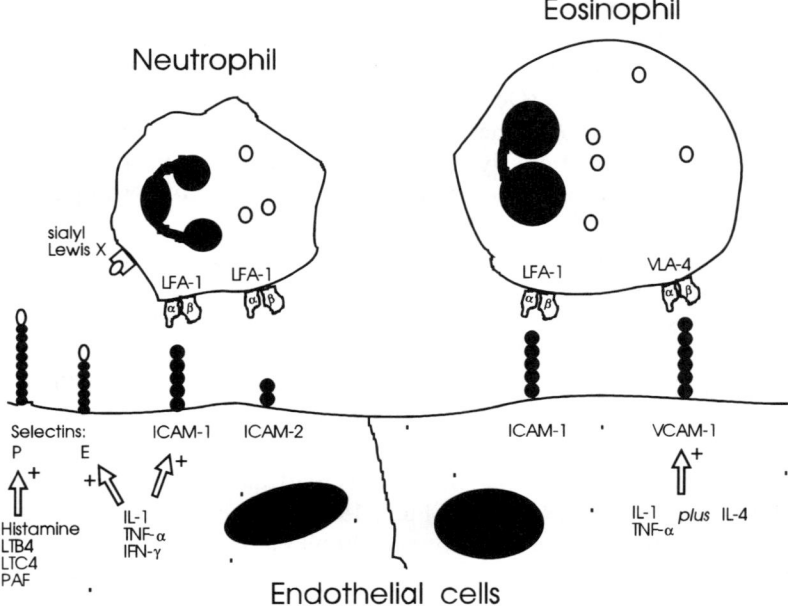

Figure 2.2 Mechanisms of leucocyte adhesion to vascular endothelium.

intercellular adhesion molecule 1 (ICAM-1) found on the endothelium (Montefort et al., 1994a). Flattening of the leucocyte on the endothelial wall is the first stage in its transendothelial migration. Endothelial expression of ICAM-1 is upregulated by the same cytokines which upregulate E-selectin. Blockade of ICAM-1 and E-selectin using neutralizing antibodies attenuates eosinophil and neutrophil influx in sensitized challenged monkeys, and reduces the LAR and the associated increase in responsiveness (Gundel et al., 1991). ICAM-1 and E-selectin are also upregulated on the endothelium following segmental allergen challenge in man, and this is associated with influx of LFA-1$^+$ neutrophils and eosinophils (Montefort et al., 1994a). Soluble forms of these CAMs are found in elevated levels in the circulation of asthmatics within 28 days of an acute exacerbation (Montefort et al., 1994b), but not in the BAL fluid or circulation in stable asthma.

It is difficult to see how the ICAM-1/LFA-1 interaction could produce a specific eosinophil recruitment, as neutrophils also express LFA-1. Neutrophils do not however express the integrin very late antigen 4 (VLA-4), which is found on eosinophils and T cells and interacts with vascular cell adhesion molecule 1 (VCAM-1) on endothelium (Wegner et al., 1990). VCAM-1 is upregulated by IL-1 or TNF-α but expression persists only in combination with IL-4 (Gundel et al., 1991; Vonderheide and Springer, 1992). In the nasal mucosa, allergen challenge causes a specific eosinophil and T-cell infiltration at 24 hours which is dependent on VLA-4 expression on the leucocytes and VCAM-1 expression on the endothelium (Bentley et al., 1992). However, despite the elegance of this hypothesis to explain eosinophilia in the atopic asthmatic lung, VCAM-1 expression does not seem to be consistently upregulated in bronchial biopsies either within 6 hours or at 24 hours after allergen challenge of asthmatics (Montefort et al., 1994a), or in the circulation after acute asthma exacerbations (Montefort et al., 1994b).

2.3.4 Mast Cells as Initiators of Chronic Inflammation

A further problem is the time-scale of CAM upregulation and leucocyte influx. Insufficient time appears to be available following allergen challenge for adequate *de novo* synthesis and release of the necessary cytokines, suggesting the existence of a preformed source. Immunohistochemical analysis of bronchial mucosal biopsies has recently identified the mast cell as the source of such preformed cytokines (Bradding et al., 1992, 1994; Okayama et al., 1995). TNF-α, GM-CSF, IL-3, IL-4, IL-5, IL-6, IL-8, IL-10 and IL-13 are all stored in mast cell granules, and may be rapidly released on IgE-dependent stimulation. In the presence of stem cell factor (SCF; kit ligand), IgE-dependent stimulation of mast cells also induces increased transcription of mRNA for IL-4, IL-5 and IL-13, with IL-5 transcripts persisting for up to 72 hours (Table 2.1). Rapid release from granules of TNF-α may upregulate ICAM-1 expression, IL-4 may upregulate VCAM-1 expression, while IL-3, IL-5 and IL-8 may

Table 2.1 Cytokine profile of purified human lung mast cells

Cytokine	SCF	SCF + anti-IgE
IL-2	−	−
IL-3	−	+
IL-4	−	+
IL-5	+	+ +
IL-6	+	+
IL-8	+ +	+ +
IL-10	+	+
IL-13	+	+ +
IFN-γ	−	−
GM-CSF	+	+
TNF-α	+	+

Mast cells were purified to greater than 98 per cent by positive immunomagnetic selection using the anti-*c-kit* antibody YB5.B8 and negative selection of T cells using anti-CD2. Cells were maintained in stem cell factor (SCF) in the presence or absence of anti-IgE. RT-PCR analysis showed that transcripts for IL-5, IL-6, IL-8, IL-10, IL-13, GM-CSF and TNF-α were constitutively expressed. IgE-dependent stimulation induced transcription of IL-3, IL-4, IL-5 and IL-13. (Data courtesy of Professor M. K. Church, Immunopharmacology Group, Southampton General Hospital.)

prime and chemoattract eosinophils and neutrophils. The induction of persistent IL-5 synthesis will perpetuate the initial eosinophilia over several days. Mast cell cytokines thus have central roles both in initiating and in perpetuating the inflammatory response to allergen exposure. 'Mast cell stabilizing drugs' such as sodium cromoglycate may act not only to reduce release of preformed and *de novo* bronchospastic mediators, but may also damp down chronic inflammation by preventing release of mast cell cytokines.

2.3.5 T-cell Orchestration of Chronic Asthma

The mast cell is ideally placed to initiate a rapid and robust response to allergen exposure. They are numerous on and beneath the bronchial epithelium and adjacent to bronchial smooth muscle and blood vessels. They store large quantities of bronchospastic autacoids and inflammatory cytokines and every mast cell is likely to have at least some allergen-specific IgE bound to its surface Fc_ε receptors. In contrast, very few T cells in the lung (much less than 1 per cent) are specific for any one allergen and cytokine release upon activation may require several hours to become significant. It seems unlikely that cytokine release from lymphocytes is responsible for the alterations in CAM expression which must take place within 2 to 3 hours of allergen exposure.

The T cell may nevertheless be a critical orchestrator of the response to allergen exposure in the long term as it is the specificity of the TCR which ultimately defines the quality or specificity of the immune response by fine-tuning the production of specific IgE by B lymphocytes (Figure 2.3). Only when equipped with this specific IgE are mast cells able to respond effectively and rapidly to allergen exposure. Allergic disease is associated with allergen-specific clones of the Th2 phenotype, expressing the cytokines of the IL-4 gene cluster on chromosome 5, including IL-4, IL-5 and IL-13, but not IL-2 or IFN-γ (Del Prete, 1992). In contrast, delayed-type hypersensitivity reactions such as tuberculosis are associated with the Th1 phenotype in which IFN-γ and IL-2 are produced, but not IL-4 or IL-5. Epitopes on allergens are recognized by dendritic cells and processed fragments are presented to T cells by an interaction involving MHC class II molecules and the TCR. In the presence of IL-4, the T cells are forced to differentiate along the Th2 pathway with further expression of cytokines of the IL-4 gene cluster. That this gene cluster on the long arm of chromosome 5 is central in allergy is supported by studies showing linkage of markers at this locus (5q31.1) with indices of atopy and asthma, particularly IgE hyperresponsiveness (Marsh *et al.*, 1994; Meyers *et*

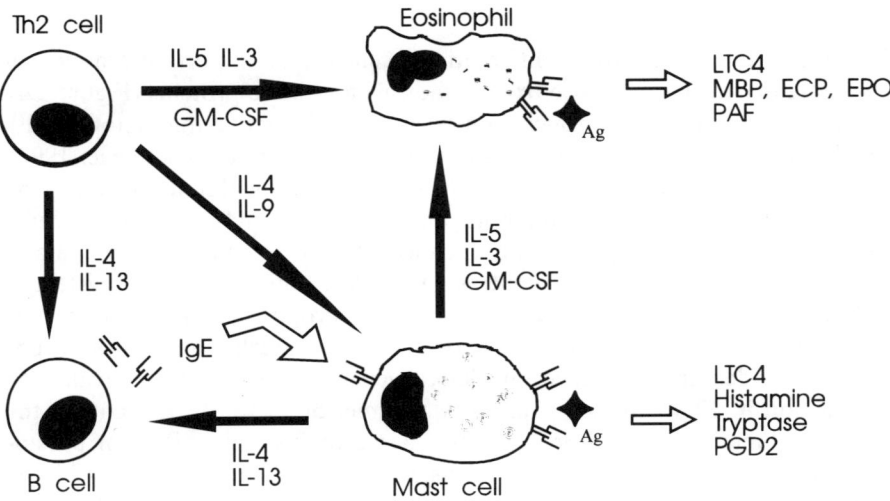

Figure 2.3 T-lymphocyte and mast cell regulation of IgE production and eosinophilia.

al., 1994). As well as IL-4, this gene cluster contains genes for eosinophilopoietic cytokines (IL-5, GM-CSF), mast cell growth factors (IL-6, IL-9), an inhibitory IFN-γ regulatory factor and receptors for β_2-adrenergic agonists and corticosteroids.

Continued expression by Th2 cells of IL-4 and IL-13 and their action at high or low affinity specific receptors on B cells induces isotype switching from IgM and IgG to IgE (Del Prete *et al.*, 1988), while IL-13 also promotes differentiation of dendritic cells in a positive feedback loop (Banchereau *et al.*, 1994). Isotype switching to IgE by IL-4 is potently inhibited by IFN-γ from Th1 cells and macrophages (Snapper and Paul, 1987). T cells in the BAL and biopsies of asthmatics are not only activated, as judged by expression of the markers CD25 (IL-2R), LFA-1, VLA-4 and HLA-DR (Azzawi *et al.*, 1990; Wilson *et al.*, 1992), but cytokine mRNA transcripts also indicate the presence of cells of the Th2 phenotype (Hamid *et al.*, 1991; Robinson *et al.*, 1992). Although allergen-specific T-cell clones are Th2-like in acute asthma, in ongoing chronic asthma the T-cell cytokine repertoire is more diverse, suggesting that the majority of T cells even in the asthmatic lung have functions other than allergen recognition (Krug *et al.*, 1996). There is a spectrum of T-cell activation with the most T-cell activation being associated with severe disease. Mild disease may be mast cell-driven, perhaps explaining the efficacy of sodium cromoglycate at this end of the disease spectrum. In chronic moderate-to-severe asthma however, the pathophysiological changes in the asthmatic lung are increasingly thought to be due to a T-cell-driven eosinophilic inflammation, which in turn leads to airway remodelling and a progressive increase in the component of irreversible airflow impairment.

2.3.6 Eosinophil Mediators

The eosinophil is a terminally differentiated cell with an important armoury of inflammatory mediators, toxic oxygen radicals and basic proteins (Figure 2.4). Eosinophils produce the bronchoconstrictor platelet-activating factor (PAF), which was once widely touted as a crucial mediator in asthma (Page, 1992), but which specific antagonists have shown to play no role (Kuitert *et al.*, 1995). However, the eosinophil is also a potent source of cysteinyl-leukotrienes which have long been thought to have a bronchoconstrictor role in acute asthma and after allergen challenge. In contrast to PAF, antagonists to cysteinyl-leukotrienes are entering clinical practice as oral prophylactic agents to reduce significantly the incidence and severity of acute exacerbations. These drugs have further demonstrated a chronic leukotriene-dependent tone, equivalent to a reduction in FEV_1 (forced expiratory volume) of 15 to 20 per cent, in the 'stable' asthmatic airway, a property not shared by histamine, PAF or prostanoids (Holgate *et al.*, 1996). This is consistent with the demonstration of higher chronic urinary excretion of LTE_4 (but not histamine) in asymptomatic stable asthmatics compared with normals (Asano *et al.*, 1995). Cysteinyl-leukotrienes

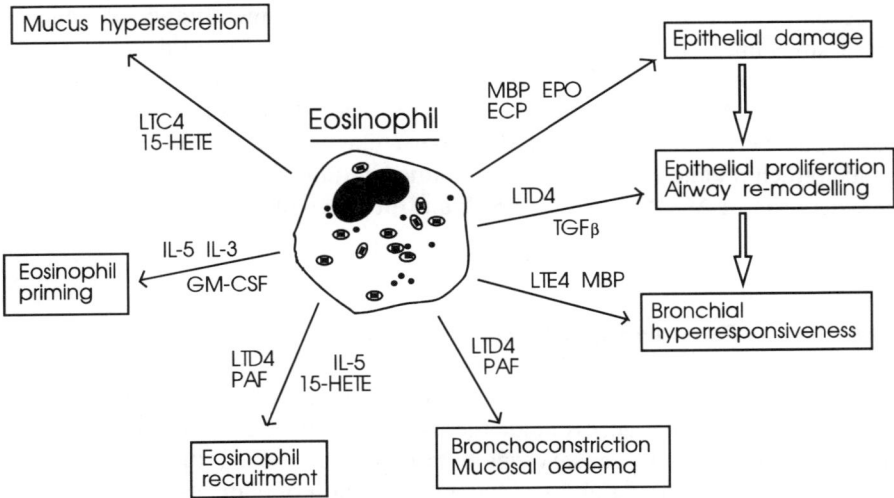

Figure 2.4 Eosinophil products implicated in the pathophysiology of asthma.

also induce bronchial hyperresponsiveness, and this may be related to the further recent recognition that they are potent and specific chemoattractants for eosinophils *in vitro* and *in vivo* (Hay et al., 1995). Leukotriene antagonists and synthesis inhibitors seem to reduce eosinophil and basophil influx after allergen challenge in man and to show significant steroid-sparing effects in severe asthma (Holgate *et al.*, 1996). This suggests a proinflammatory role for cysteinyl-leukotrienes which can be hypothesized to include a positive feedback loop of cysteinyl-leukotriene production by eosinophils causing further eosinophil influx. The eosinophil is a source of proinflammatory cytokines including IL-3, IL-5, GM-CSF and TNF-α (Kita *et al.*, 1991; Moqbel *et al.*, 1991; Costa *et al.*, 1993). These factors may also act in an autocrine or paracrine fashion to prime and chemoattract eosinophils for enhanced mediator release and enhance their own survival.

2.3.7 Eosinophil Basic Proteins and the Bronchial Epithelium

In addition to the synthesis of bronchoconstrictor mediators and cytokines, the eosinophil is the cell most closely implicated in bronchial epithelial damage in asthma. Clumps of shed epithelial cells (Creola bodies) were described in the sputum of asthmatics after an acute exacerbation by Naylor (1962) and stripping of the pseudostratified ciliated epithelium down to the basal cell layer occurs over large areas in the post-mortem lung of status asthmaticus patients. In mild-to-moderate asthma, the epithelial cleavage occurs along the line of desmosomes in the plane between the columnar and basal cells (Montefort *et al.*, 1992a, 1992b), while the basal cells are more firmly attached by a line of hemidesmosomes to the basement membrane. Damage to the desmosomes may occur by the release of metalloendopeptidases (e.g. gelatinase A) from eosinophils, or by the action of the arginine-rich basic proteins: eosinophil cationic protein (ECP), major basic protein (MBP) or eosinophil peroxidase (EPO). MBP in particular is highly cytotoxic to bronchial epithelium and is elevated in asthmatic sputum (Frigas and Gleich, 1986), and deposits of MBP and ECP are found in areas of epithelial loss in postmortem asthmatic lung.

2.3.8 Airway Remodelling and Repair in Chronic Asthma

The loss of bronchial epithelium and its ciliary layer will impair mucus clearance and allow greater access of allergens and noxious stimuli to the bronchial smooth muscle and to sensory nerve receptors. However, the epithelium is much more than a simple physical barrier. Secretion of platelet-derived growth factor (PDGF), endothelin-1 (ET-1) and basic fibroblast growth factor (bFGF) by damaged epithelium leads to proliferation of subepithelial myofibroblasts. These myofibroblasts in turn secrete types III and V collagen which are laid down beneath the basement membrane giving rise to the appearance of a 'thickened basement membrane', a typical pathological feature of the asthmatic lung (Roche et al., 1989). The number of subepithelial myofibroblasts has been shown to correlate closely with the degree of collagen deposition (Brewster et al., 1990). Myofibroblasts also secrete the cytokines GM-CSF, IL-6, IL-8 and SCF and thus may regulate eosinophil and mast cell activity in mucosal inflammation.

Chronic inflammation in the asthmatic lung is associated with proliferative and repair processes which may lead to progressive fibrosis and a reduced reversibility of airflow impairment. These processes include not only widespread destruction of bronchial epithelium, but also the proliferation and activation of fibroblasts, hypertrophy and hyperplasia of bronchial smooth muscle, and proliferation of neuropeptide-secreting nerves. Understanding the way in which cytokines released by epithelium, macrophages and fibroblasts, including TNF-α, IL-1, platelet-derived growth factor (PDGF) and various nerve growth factors, may regulate this progression may be of fundamental importance in combating inflammatory damage in severe chronic asthma.

2.4 Trends in Asthma Treatment

2.4.1 The Emphasis on Anti-inflammatory Therapy

Effective treatment of asthma is currently based on two main groups of asthma drugs, the bronchodilators such as β_2-agonists, theophylline and ipratropium bromide and the anti-inflammatory agents, particularly glucocorticosteroids and the cromones. However, underdiagnosis of asthma and of its severity is common, and misunderstanding of asthma drugs is widespread, with patients often failing to appreciate the difference between bronchodilators used for symptomatic relief and corticosteroids used prophylactically. Compliance with inhaled medication is often lamentable due to poor inhaler technique. Many patients and some physicians are wary of inhaled corticosteroids, fearing side-effects associated with oral steroid use. Although side-effects include suppression of the hypothalamo–pituitary axis, osteoporosis, skin thinning and bruising, cataracts, weight gain and growth suppression in children, these become apparent only rarely even at high daily doses of inhaled steroids (Geddes, 1992).

The problems of underdiagnosis of asthma severity and poor drug compliance have been addressed by recent guidelines on the treatment of chronic asthma (British Thoracic Society, 1993; National Institutes of Health, 1995). The value of avoidance of asthma triggers such as allergen, aspirin, or occupational sensitizing chemicals is recognized, but avoidance of some may be difficult (house dust mite allergen) or impossible (viral respiratory tract infections), or impose restrictions on lifestyle (exercise/cold air) which are less acceptable than appropriate medication. Guidelines stress a stepwise approach to asthma of varying severity; increased understanding of the inflammatory processes underlying asthma, and of the progressive inflammatory damage in the chronic asthmatic lung, have led to an emphasis on the early introduction of anti-inflammatory therapy in any patient requiring more than occasional use of their β_2-agonist inhaler, and in those waking regularly at night with asthma symptoms. Combined with better diagnosis of the severity of asthma, and better patient education to improve compliance, this approach is likely to make a significant impact on asthma morbidity and mortality.

2.4.2 New Drugs

The principal asthma therapies have all been in clinical practice since the early 1970s and before, yet their mechanisms and potential hazards in many cases remain obscure. The trend towards inhaled corticosteroids continues to raise concerns about their long-term side-effects. This has fuelled efforts to develop 'safer' anti-inflammatory therapies, despite the advent of fluticasone, which is 100 per cent first-pass metabolized in the liver and has fewer systemic effects (Harding, 1990). Suggestions that the long-acting β_2-agonists (salmeterol, formoterol) may be anti-inflammatory appear to be unfounded, but the possibility of an anti-inflammatory effect of theophylline (Ward et al., 1993) has accelerated development of selective phosphodiesterase (PDE IV) inhibitors, which may have reduced side-effects and a better therapeutic index than theophylline itself. The immunosuppressants, such as cyclosporin A which prevents expression of IL-2 and IL-2R in T cells, are limited by toxicity to a small minority of very severe corticosteroid-dependent asthmatics.

Some new compounds have targetted particular families of inflammatory mediators, including PAF and thromboxane, but most have proved disappointing (Barnes, 1993). The exceptions are the leukotriene antagonists (pranlukast, zafirlukast, verlukast) and synthesis inhibitors (zileuton), which have passed through clinical trials and some of which are now in the process of being licensed for widespread use. These drugs are likely to be used across the spectrum of asthma to reduce the use or the dose of corticosteroids, and in children who may not tolerate theophylline well (Sampson and Costello, 1995; Holgate et al., 1996). Indications that they reduce inflammatory cell influx into the lung are being followed up.

A long way behind in the line of development are antagonists to IL-4 and IL-5, which might be hoped to combat the IgE overproduction and eosinophilia which are central to allergic asthma. Although a highly promising approach, it is possible that antagonism of IL-4 may only reduce the drive to overproduction of total IgE, but not that of allergen-specific IgE. Interleukin-5 antagonists may reduce eosinophil proliferation, but experiments in IL-5 transgenic mice suggest that despite massive eosinophilia in the lung and other organs, these mice are otherwise healthy. The presence of eosinophils in the bronchial mucosa does not in itself equate with asthma, and much research is still needed into the combination of factors causing chronic inflammatory damage. Other novel therapeutic avenues being explored include 'humanized' monoclonal antibodies against IgE, antagonists of VLA-4 and a range of peptides and other agents capable of manipulating T-cell orchestration of the inflammatory response.

References

AMERICAN THORACIC SOCIETY (1962) Committee on diagnostic standards, definitions and classification of chronic bronchitis, asthma and pulmonary emphysema. *American Review of Respiratory Disease*, **85**, 762–768.

ARM, J. P. & LEE, T. H. (1993) Sulphidopeptide leukotrienes in asthma. *Clinical Science*, **84**, 501–510.

ARSHAD, S. H., MATTHEWS, S., GRANT, C. & HIDE, D. W. (1993) Effect of allergen avoidance on development of allergic disorders in infancy. *Lancet*, **339**, 1493–1497.

ASANO, K., LILLY, C. M., O'DONNELL, W. J., ISRAEL, E., FISCHER, A., RANSIL, B. J. & DRAZEN, J. M. (1995) Diurnal variation of urinary leukotriene E_4 and histamine excretion rates in normal subjects and patients with mild-to-moderate asthma. *Journal of Allergy and Clinical Immunology*, **96**, 643–651.

AZZAWI, M., BRADLEY, B., JEFFREY, P. K., FREW, A. J., WARDLAW, A. J., KNOWLES, G., ASSOUFI, B., COLLINS, J. V., DURHAM, S. & KAY, A. B. (1990) Identification of activated T lymphocytes and eosinophils in bronchial biopsies in stable atopic asthma. *American Review of Respiratory Disease*, **142**, 1407–1413.

BANCHEREAU, J., BRIERE, F., LIU, Y. J. & ROUSSET, F. (1994) Molecular control of B lymphocyte growth and differentiation. *Stem Cells*, **12**, 278–288.

BARNES, N. C. (1993) New developments in the treatment of asthma and chronic obstructive pulmonary disease. *Respiratory Medicine*, **87** (Suppl. B), 53–56.

BEASLEY, R., ROCHE, W. R., ROBERTS, J. A. & HOLGATE, S. T. (1989) Cellular events in the bronchi in mild asthma and after bronchial provocation. *American Review of Respiratory Disease*, **139**, 806–817.

BENTLEY, A. M., DURHAM, S. R., ROBINSON, D. R., HAMID, Q., KAY, A. B. & DURHAM, S. R. (1992) Expression of the endothelial leucocyte adhesion molecules ICAM-1, E-selectin and UCAM-1 in the bronchial mucosa in steady state asthma and allergen-induced asthma. *Thorax*, **47**, 852P (Abstract).

BENTLEY, A. M., MENG, Q., ROBINSON, D. S., HAMID, Q., KAY, A. B. & DURHAM, S. R. (1993) Increases in activated T lymphocytes, eosinophils and cytokine mRNA expression for interleukin-5 and granulocyte/macrophage colony-stimulating factor in bronchial biopsies after allergen inhalation challenge in atopic asthmatics. *American Journal of Respiratory Cell and Molecular Biology*, **8**, 35–42.

BLACKLEY, C. H. (1873) *Experimental Researches on the Causes and Nature of Catarrhus Aestivus (hay fever or hay-asthma)*. London: Baillière Tindall and Cox.

BRADDING, P., FEATHER, I. H., HOWARTH, P. H., MUELLER, R., ROBERTS, J. A. & BRITTEN, K. (1992) Interleukin 4 is localized to and released by human mast cells. *Journal of Experimental Medicine*, **176**, 1381–1386.

BRADDING, P., ROBERTS, J. A., BRITTEN, K. M., MONTEFORT, S., DJUKANOVIĆ, R., MUELLER, R., HEUSSER, C. H., HOWARTH, P. H. & HOLGATE, S. T. (1994) Interleukin-4, -5 and -6 and tumor necrosis factor-alpha in normal and asthmatic airways: evidence for the human mast cell as a source of these cytokines. *American Journal of Respiratory Cell and Molecular Biology*, **10**, 471–480.

BREWSTER, C. E. P., HOWARTH, P. H., DJUKANOVIĆ, R., WILSON, J., HOLGATE, S. T. & ROCHE, W. R. (1990) Myofibroblasts and sub-epithelial fibrosis in bronchial asthma. *American Journal of Respiratory Cell and Molecular Biology*, **3**, 507–511.

BRITISH THORACIC SOCIETY (1993) Guidelines on the management of asthma. *Thorax*, **48** (Suppl.), S1–S24.

BROWN, P. H., CROMPTON, G. K. & GREENING, A. P. (1991) Proinflammatory cytokines in acute asthma. *Lancet*, **338**, 590–593.

CALHOUN, W. J., SEDGWICK, J. & BUSSE, W. W. (1991) The role of eosinophils in the pathophysiology of asthma. *Annals of the New York Academy of Sciences*, **629**, 62–72.

COCKCROFT, D. W. & MURDOCK, K. Y. (1987) Changes in bronchial responsiveness to histamine at intervals after allergen challenge. *Thorax*, **42**, 302–308.

COOKSON, W. O. C. M. (1993) Genetic aspects of atopy. *Monographs in Allergy*, **31**, 171–189.

COOKSON, W. O. C. M., CRADDOCK, C. F., BENSON, M. K. & DURHAM, S. R. (1989) Falls in peripheral eosinophil counts parallel the late asthmatic response. *American Review of Respiratory Disease*, **139**, 458–462.

COSTA, J. J., MATOSSIAN, K., RESNICK, M. B., BEIL, W. J., WANG, D. T., GORDON, J. R., DVORAK, A. M., WELLER, P. F. & GALLI, S. J. (1993) Human eosinophils can express the cytokines tumour necrosis factor-α and macrophage inflammatory protein-1α. *Journal of Clinical Investigation*, **91**, 2673–2684.

DEL PRETE, G. (1992) Human Th1 and Th2 lymphocytes: their role in the pathophysiology of atopy. *Allergy*, **47**, 450–455.

DEL PRETE, G. F., MAGGI, E., PARRONCHI, P., CHRETIEN, I., TIRI, A., MACCHIA, D., RICCI, M., BANCHEREAU, J., DE VRIES, J. & ROMAGNANI, S. (1988) IL-4 is an essential co-factor for the IgE synthesis induced *in vitro* by human T cell clones and their supernatants. *Journal of Immunology*, **140**, 4193–4197.

DE MONCHY, J. G. R., KAUFFMAN, H. F., VENGE, P., KOETER, G. H., JANSEN, H. M., SLUITER, H. J. & DE VRIES, K. (1985) Bronchoalveolar lavage eosinophilia during allergen-induced late asthmatic reactions. *American Review of Respiratory Disease*, **131**, 373–376.

DJUKANOVIĆ, R., WILSON, J. W., BRITTEN, K. M., WILSON, S. J., WALLS, A. F., ROCHE, W. R., HOWARTH, P. H. & HOLGATE, S. T. (1990) Quantitation of mast cells and eosinophils in the bronchial mucosa of symptomatic atopic asthmatics and healthy control subjects using immunohistochemistry. *American Review of Respiratory Disease*, **142**, 863–871.

DUNNILL, M. S. (1960) The pathology of asthma with special reference to changes in the bronchial mucosa. *Journal of Clinical Pathology*, **13**, 224–225.

FREW, A. J., TERAN, L. M., MADDEN, J. M., TREFILIEFF, A., ST. PIERRE, J., SEMPER, A., CARROLL, M. P. & HOLGATE, S. T. (1995) Cellular changes 24 hours after endobronchial allergen challenge in asthma. *International Archives of Allergy and Applied Immunology*, **107**, 376–377.

FRIGAS, E. & GLEICH, G. J. (1986) The eosinophil and the pathophysiology of asthma. *Journal of Allergy and Clinical Immunology*, **77**, 527–537.

GEDDES, D. (1992) Inhaled corticosteroids: benefits and risks. *Thorax*, **47**, 404–407.

GODFREY, K. M., BARKER, D. J. P. & OSMOND, C. (1994) Disproportionate fetal growth and raised IgE concentration in adult life. *Clinical and Experimental Allergy*, **24**, 641–648.

GUNDEL, R. H., WEGNER, C. D., TORCELLINI, C. A., CLARKE, C. C., HAYNES, N., ROTHLEIN, R., SMITH, C. W. & LETTS, L. G. (1991) ELAM-1 mediates antigen-induced acute airway inflammation and late phase obstruction in monkeys. *Journal of Clinical Investigation*, **88**, 1407–1411.

HAMID, Q., AZZAWI, M., YING, S., MOQBEL, R., WARDLAW, A. J., CORRIGAN, C. J., BRADLEY, B., DURHAM, S. R., COLLINS, J. V. & JEFFERY, P. K. (1991) Expression of mRNA for interleukin-5 in mucosal bronchial biopsies from asthmatics. *Journal of Clinical Investigation*, **87**, 1541–1546.

HARDING, S. M. (1990) The human pharmacology of fluticasone propionate. *Respiratory Medicine*, **84**, 25–29.

HAY, D. W. P., TORPHY, T. J. & UNDEM, B. J. (1995) Cysteinyl leukotrienes in asthma: old mediators up to new tricks. *Trends in Pharmacological Science*, **16**, 304–309.

HOLGATE, S. T. & CHURCH, M. K. (1993) *Allergy*. London: Gower Medical Publishing.

HOLGATE, S. T., BRADDING, P. & SAMPSON, A. P. (1996) Leukotriene antagonists and synthesis inhibitors: new directions in asthma therapy. *Journal of Allergy and Clinical Immunology* (in press).

JEFFERY, P. K., WARDLAW, A. J., NELSON, F. C., COLLINS, J. V. & KAY, A. B. (1989) Bronchial biopsies in asthma: an ultrastructural quantification study

and correlation with hyperreactivity. *American Review of Respiratory Disease*, **140**, 1745–1753.
KITA, H., OHNISHI, T., OKUBO, Y., WEILER, D., ABRAMS, J. J. & GLEICH, G. J. (1991) Granulocyte/macrophage colony-stimulating factor and interleukin-3 release from human peripheral blood eosinophils and neutrophils. *Journal of Experimental Medicine*, **174**, 745–758.
KRUG, N., MADDEN, J., REDINGTON, A. E., HOWARTH, P., HOLGATE, S. T., DJUKANOVIĆ, R. & FREW, A. J. (1996) Intracellular cytokine staining of T cells obtained from human asthmatic airways. *American Journal of Respiratory and Critical Care Medicine* (in press).
KUITERT, L. M., ANGUS, R. M., BARNES, N. C., BARNES, P. J., BONE, M. F., CHUNG, K. F., FAIRFAX, A. J., HIGENBOTHAM, T. W., O'CONNOR, B. J. & PIOTROWSKA, B. (1995) Effect of a novel potent platelet-activating factor antagonist, modipafant, in clinical asthma. *American Journal of Respiratory and Critical Care Medicine*, **151**, 1331–1335.
MARSH, D. G. & MEYERS, D. A. (1992) A major gene for allergy: fact or fancy? *Nature Genetics*, **2**, 252–254.
MARSH, D. G., NEELY, J. D., BREAZEALE, J., GHOSH, B., FREIDHOFF, L. R., EHRLICH-KAUTSKY, E., SCHOU, C., KRISHNASWAMY, G. & BEATY, T. H. (1994) Linkage analysis of IL-4 and other chromosome 5q31.1 markers and total serum immunoglobulin E concentrations. *Science*, **264**, 1152–1156.
METZGER, W. J., ZAVALA, D., RICHERSON, H. B., MOSELEY, P., IWAMOTO, P., MONICK, M., SJOERDSMA, K. & HUNNINGHAKE, G. W. (1987) Local allergen challenge and bronchoalveolar lavage of allergic asthmatic lungs: description of the model and local airway inflammation. *American Review of Respiratory Disease*, **135**, 433–440.
MEYERS, D. A., POSTMA, D. S., PANHUYSEN, C. I. M., XU, J., AMELUNG, P. J., LEVITT, R. C. & BLEECKER, E. R. (1994) Evidence for a locus regulating total serum IgE levels mapping to chromosome 5. *Genomics*, **23**, 464–470.
MOFFATT, M. F., HILL, M. R., CORNELIS, F., SCHOU, C., FAUX, J. A., YOUNG, R. P., JAMES, A. L., RYAN, G., LE SOUEF, P. & MUSK, A. W. (1994) Genetic linkage on T-cell receptor α/δ complex to specific IgE responses. *Lancet*, **343**, 1579–1599.
MONTEFORT, S., HERBERT, C. A., ROBINSON, C. & HOLGATE, S. T. (1992a) The bronchial epithelium as a target for inflammatory attack in asthma. *Clinical and Experimental Allergy*, **22**, 511–520.
MONTEFORT, S., ROBERTS, J. A., BEASLEY, R., HOLGATE, S. T. & ROCHE, W. R. (1992b) The site of disruption of the bronchial epithelium in asthmatic and non-asthmatic subjects. *Thorax*, **47**, 499–503.
MONTEFORT, S., GRATZIOU, C., GOULDING, D., POLOSA, R., HASKARD, D. O., HOWARTH, P. H., HOLGATE, S. T. & CARROLL, M. P. (1994a) Bronchial biopsy evidence for leucocyte infiltration and upregulation of leucocyte-endothelial cell adhesion molecules 6 hours after local allergen challenge of sensitized asthmatic airways. *Journal of Clinical Investigation*, **93**, 1411–1421.
MONTEFORT, S., LAI, C. K., KAPAHI, P., LEUNG, J., LAI, K.N., CHAN, H. S., HASKARD, D. O., HOWARTH, P. H. & HOLGATE, S. T. (1994b) Circulating adhesion molecules in asthma. *American Journal of Respiratory and Critical Care Medicine*, **149**, 1149–1152.

MOQBEL, R., HAMID, Q., YING, S., BARKANS, J., HARTNELL, A., TSICOPOULOS, A., WARDLOW, A. J. & KAY, A. B. (1991) Expression of mRNA and immunoreactivity for the granulocyte/macrophage colony-stimulating factor (GM-CSF) in activated human eosinophils. *Journal of Experimental Medicine*, **174**, 749–753.

NATIONAL INSTITUTES OF HEALTH (1995) *Global Initiative for Asthma*. Global strategy for asthma management and prevention NHBLI/WHO Workshop, NIH Publication 95-3659.

NAYLOR, B. (1962) The shedding of the mucosa of the bronchial tree in asthma. *Thorax*, **17**, 69–72.

OKAYAMA, Y., PETIT FRERE, C., KASSEL, O., SEMPER, A., QUINT, D., TUNON DE LARA, M. J., BRADDING, P., HOLGATE, S. T. & CHURCH, M. K. (1995) IgE-dependent expression of mRNA for IL-4 and IL-5 in human lung mast cells. *Journal of Immunology*, **155**, 1796–1808.

PAGE, C. P. (1992) Mechanisms of hyperresponsiveness: platelet-activating factor. *American Review of Respiratory Disease*, **145**, S31–S33.

PEPYS, J. (1973) Disodium cromoglycate in clinical and experimental asthma, in AUSTEN, K. F. & LICHTENSTEIN, L. M. (Eds). *Asthma: physiology, immunopharmacology and treatment*, pp. 279–294. London: Academic Press.

PORTER, R. & BIRCH, J. (1971) *Identification of asthma*. Ciba Foundation Study Group No. 38. Edinburgh: Churchill Livingstone.

ROBINSON, D. S., HAMID, Q., YING, S., TSICOPOULOS, A., BARKANS, J., BENTLEY, A. M., CORRIGAN, C., DURHAM, S. R. & KAY, A. B. (1992) Predominant TH2-like bronchoalveolar T-lymphocyte population in atopic asthma. *New England Journal of Medicine*, **326**, 298–304.

ROCHE, W. R., BEASLEY, R., WILLIAMS, J. H. & HOLGATE, S. T. (1989) Sub-epithelial fibrosis in the bronchi of asthmatics. *Lancet*, **i**, 520–524.

ROYAL COLLEGE OF PHYSICIANS (1992) *Smoking and the Young*, pp. 1–23, London.

RUSSI, J. C., DELFRARO, A., BORTHAGARAY, M. D., VELAZQUEZ, B., GARCIA-BARRENO, B. & HORTAL, M. (1993) Evaluation of immunoglobulin E-specific antibodies and viral antigens in nasopharyngeal secretions of children with respiratory syncytial virus infections. *Journal of Clinical Microbiology*, **31**, 819–823.

SALTER, H. H. (1868) *On asthma: its pathology and treatment*, 2nd edn. London: Churchill.

SAMPSON, A. P. & COSTELLO, J. F. (1995) Treatment guidelines for asthma: where do leukotriene receptor antagonists fit in? *The Pharmaceutical Journal*, **255**, 26–30.

SANDFORD, A. J., SHIRAKAWA, T., MOFFATT, M. F., DANIELS, S. E., RA, C., FAUX, J. A., YOUNG, R. P., NAKAMURA, Y., LATHROP, G. M. & COOKSON, W. O. (1993) Localisation of atopy and beta subunit of high-affinity IgE receptor (Fc$_\varepsilon$RI) on chromosome 11q. *Lancet*, **341**, 332–334.

SEHMI, R., WARDLAW, A. J., CROMWELL, O., KURIHARA, K., WALTMANN, P. & KAY, A. B. (1992) Interleukin-5 selectively enhances the chemotactic response of eosinophils obtained from normal but not eosinophilic subjects. *Blood*, **79**, 2952–2959.

SHUTE, J. (1994) Interleukin-8 is a potent eosinophil chemoattractant. *Clinical and Experimental Allergy*, **24**, 203–206.

SIMONS, F. E. R. (1994) *Ancestors of Allergy*. New York: Global Medical Communications Ltd.

SMITH, L. & MCFADDEN, E. R. (1995) Bronchial hyperreactivity revisited. *Annals of Allergy, Asthma and Immunology*, **74**, 454–469.

SNAPPER, C. M. & PAUL, W. E. (1987) Interferon-gamma and B cell stimulatory factor-1 reciprocally regulate Ig isotype production. *Science*, **236**, 944–947.

SPORIK, R., HOLGATE, S., PLATTS-MILLS, T. A. E. & COGSWELL, J. J. (1990) Exposure to house-dust mite allergen (*Der p* 1) and the development of asthma in childhood. *New England Journal of Medicine*, **323**, 502–507.

STEWART, G. A. (1994) The molecular biology of allergens, in HOLGATE, S. T. & BUSSE, W. (Eds). *The Mechanisms of Asthma and Rhinitis*. Oxford: Blackwells.

SUR, S., GLEICH, G. J., SWANSON, M. C., BARTEMES, K. R. & BROIDE, D. H. (1995) Eosinophilic inflammation is associated with elevation of interleukin-5 in the airways of patients with spontaneous symptomatic asthma. *Journal of Allergy and Clinical Immunology*, **96**, 661–668.

VONDERHEIDE, R. H. & SPRINGER, T.A. (1992) Lymphocyte adhesion through the very late antigen 4: evidence for a novel binding site in the alternatively spliced domain of VCAM-1 and an additional $\alpha 4$ integrin counter receptor on stimulated endothelium. *Journal of Experimental Medicine*, **175**, 1433–1442.

WALLS, A. F., HE, S., TERAN, L., BUCKLEY, M. G., KI-SUCK, J., HOLGATE, S. T., SHUTE. J. K. & CAIRNS, J. A. (1995) Granulocyte recruitment by human mast cell tryptase. *International Archives of Allergy and Immunology*, **107**, 372–373.

WARD, A. J., MCKENNIFF, M., EVANS, J. M., PAGE, C. P. & COSTELLO, J. F. (1993) Theophylline – an immunomodulatory role in asthma? *American Review of Respiratory Disease*, **147**, 518–523.

WARNER, J. A., MILES, E. A., JONES, A. C., QUINT, D. J., COLWELL, B. M. & WARNER, J. O. (1994) Is deficiency of interferon gamma production by allergen triggered cord blood cells a predictor of atopic asthma? *Clinical and Experimental Allergy*, **24**, 423–430.

WEGNER, C. D., GUNDEL, R. H., REILLY, P., HAYNES, N., LETTS, L. G. & ROTHLEIN, R. (1990) ICAM-1 in the pathogenesis of asthma. *Science*, **247**, 416–418.

WILSON, J. W., DJUKANOVIĆ, R., HOWARTH, P. H. & HOLGATE, S. T. (1992) Lymphocyte activation in bronchoalveolar lavage and peripheral blood in atopic asthma. *American Review of Respiratory Disease*, **145**, 958–960.

chemicals emanates from index case reports, cross-sectional surveys or, rarely, prospective epidemiologic studies in the workplace. The workplace is in many respects an ideal epidemiological model for studying respiratory allergy since the causative agent(s) and source(s) of exposure can often be identified and correlated with the workers' exposure and clinical symptoms. In addition, it is often possible to investigate specific immune responses associated with exposure and disease. The SWORD (Surveillance of Work Related and Occupational Respiratory Disease) project, initiated in the United Kingdom in 1989, is an excellent example of an epidemiological model which was established to monitor the frequency of occupational respiratory disorders. SWORD was also designed to promote early recognition of these disorders and facilitate investigation and prevention of work-related respiratory disease (Meredith et al., 1991). Crude annual incidence rates of the most common occupational respiratory diseases reported by SWORD are summarized in Table 3.1 (Meredith et al., 1991). The most common work-related respiratory disease was found to be asthma with an estimated annual rate of 22 cases per million individuals, followed by malignant mesothelioma and pneumoconiosis. Diisocyanates were identified as the most common cause of occupational

Table 3.1 SWORD crude annual rates by diagnostic category

Diagnostic category	No. (%)	Rate/million*
Inhalation accidents	72 (3–4)	3
Acute pulmonary oedema	15	
Other	57	
Allergic alveolitis	133 (6–3)	5
Asthma	554 (26–4)	22
Building related illness	217 (10–3)	9
Byssinosis	23 (1–1)	1
Infectious diseases	100 (4–8)	4
Lung cancer (with pulmonary fibrosis)	51 (2–4)	2
Malignant mesothelioma	340 (16–2)	13
Pneumoconiosis	322 (15–3)	13
Benign pleural disease	221 (10–5)	9
Other diagnoses	68 (3–2)	3
Bronchitis/emphysema	25	
Lung cancer	12	
Other	26	
Unspecified	5	
Total	2101 (100)	84

* Based on the 1988 Labour Force Survey estimate of the United Kingdom working population from: Meredith, S. K., Taylor, V. M. and McDonald, J. C. (1991) Occupational respiratory disease in the United Kingdom 1989: a report to the British Thoracic Society and the Society of Occupational Medicine by the SWORD project group. *British Journal of Industrial Medicine*, **48**, 292–298.

asthma (Meredith *et al.*, 1991). A surveillance programme initiated in Finland found a similar incidence of occupational asthma but a higher rate of hypersensitivity pneumonitis (farmer's lung) compared with the SWORD programme (Keskinen, 1983). Attempts have been made to compile more information about work-related respiratory diseases in the United States. The Sentinel Health Notification System for Occupational Risks (SENSOR) was established in several states to determine the incidence of the various occupational respiratory disorders. Preliminary results of SENSOR indicate that occupational asthma (OA) is the most commonly reported respiratory disease encountered by physicians (Seligman and Bernstein, 1993). In the early 1980s, the Occupational Safety Health and Administration (OSHA) established a large, computerized database called the United States National Exposure Survey Data Base (NOES) which has been compiling information pertaining to all reports of work-related exposures (Seta *et al.*, 1993). Although NOES was originally designed as a reference guide to determine the number of workers potentially exposed to various chemicals or physical and biological agents, it is also helpful in gathering prevalence statistics pertaining to chemically induced respiratory diseases and their specific causes (Seta *et al.*, 1993). Unfortunately, none of these epidemiologic models provide information regarding the natural course of the disease.

3.3 Risk Factors

One of the advantages of a well-designed epidemiologic study is that it allows determination of risk factors associated with the development of respiratory disease. Little is known about risk factors for developing most types of respiratory diseases, because of the difficulty in arranging long-term immunosurveillance programmes in specific industries. For OA, it is known that the degree of exposure to a sensitizing agent increases a worker's risk for sensitization and development of asthma. This has been demonstrated for workers exposed to acid anhydrides, diisocyanates and colophony fumes (Burge *et al.*, 1981; Becklake, 1993; Butcher *et al.*, 1993). Atopy has been demonstrated to be an important risk factor for development of OA in workers exposed to high-molecular-weight (HMW $> 5 \times 10^3$ kD) compounds but thus far a similar relationship has been demonstrated for only a few low-molecular-weight chemical agents (LMW $< 1 \times 10^3$ kD) (i.e. acid anhydride compounds and platinum metallic salts) (Becklake, 1993). Cigarette smoking is a variable risk factor among workers who develop OA (Becklake and Lalloo, 1990). For instance, some investigators have demonstrated that cigarette smoking in conjunction with atopy resulted in a higher prevalence of sensitization in workers exposed to tetrachlorophthalic anhydride (TCPA) compared with exposed, non-atopic non-smokers (Venables *et al.*, 1985). However, a subsequent study did not confirm these findings in workers exposed to TCPA or other acid anhydride agents (Liss *et al.*, 1993). Chan-Yeung found a negative correlation between cigarette smoking and red cedar-induced OA, indicating that non-smokers were actually more susceptible to developing this disease (Chan-Yeung, 1982). Non-specific bronchial hyperresponsiveness (NSBH) has also been evaluated as a risk factor for developing occupational asthma. Prospective data in the western red cedar industry indicate that NSBH occurs as the result of chemical exposure and/or allergic sensitization and does not appear to be a pre-existing risk factor for developing OA. In general, gender, age and race have not been determined to be risk factors for workers developing work-related respiratory diseases (Becklake, 1993).

Recently, it has been shown that workers who develop chronic berylliosis have a disproportionally higher incidence of the HLA profile – Dpb_1, GLU 69 (Richeldi et al., 1993). Similarly, workers more susceptible to developing diisocyanate-OA expressed the HLA allele DQB1*503 whereas workers who expressed HLA-DQB1*501 were less susceptible to this disease (Bignon et al., 1994). Whether these HLA phenotypes prove to be reliable markers that can be used to identify workers at increased risk for developing disease will require further analysis in larger populations of exposed symptomatic and asymptomatic workers.

3.4 Mechanisms of Occupational Respiratory Diseases

Since occupational asthma is the most common work-related respiratory disease, more is known about the mechanisms of specific causative agents. OA has been described in a number of different work environments including the pharmaceutical, textile, automobile manufacturing, food processing, hair care, printing, farming, smelting, soldering, metal plating, tanning and brewing industries. Occupational asthma is defined as 'a disease characterized by variable airflow limitation and/or airway hyperresponsiveness due to causes and conditions attributable to a particular occupational environment and not to stimuli encountered outside the workplace' (Bernstein, IL et al., 1993). Subcategories of occupational asthma are defined by the presence or absence of a latency period. In the presence of a defined latency period, the causes of occupational asthma can in most instances be accounted for by an underlying immunologic mechanism. Those conditions without a latency period are termed irritant-induced asthma, originally described as 'reactive airways dysfunction syndrome' (RADS) (Bernstein, IL et al., 1993).

The classical presentation of OA with a latency period is characterized by an asymptomatic 'incubation' period during which sensitization to a substance in the work environment occurs. When clinical expression of asthma symptoms finally ensue, they are elicited by low concentrations of the causative agent and they usually improve or disappear away from the work site. Symptoms might be most apparent on a Monday morning when a patient returns to work after being away from the workplace for the weekend. In general, severity of the asthma symptoms correlates well with the intensity and length of exposure to the inciting agent (Chan-Yeung, 1993).

Irritant-induced asthma (RADS) occurs in the absence of a latency period, after either a single exposure to toxic concentrations or repetitive exposures to lower concentrations of gases, strong acids and alkalis, organic solvents or wood smoke (Brooks et al., 1985). Such exposures often result in NSBH that may persist for months or years (Brooks and Bernstein, 1993).

Respiratory syndromes such as potroom asthma, which occurs in aluminium smelting workers, or byssinosis, which occurs in textile workers exposed to a variety of textile dusts such as cotton, flax, hemp or jute, should only be characterized as 'asthma-like' because of the absence of long-term NSBH and tissue eosinophils (Bernstein, IL et al., 1993).

The aetiologic agents of OA have typically been classified according to molecular weight and biological source (i.e. microbial; animal, bird or arthropod; plant; or chemicals). High-molecular-weight substances are defined as having a molecular weight greater than 5×10^3 kD whereas LMW substances have a molecular weight less than 1×10^3 kD (Bernstein, DI, 1992). This chapter will focus on LMW causes which include primarily chemical and pharmacologic agents.

The underlying mechanisms associated with induction of asthma by LMW agents are complex and only well defined for a few agents. Low-molecular-weight agents are classified on the basis of whether they induce OA through immunologic or non-immunologic mechanisms (Table 3.2). Low-molecular-weight agents are unique in that they only acquire immunogenicity after linkage with an endogenous protein carrier. Such multivalent hapten–protein conjugates are then capable of inducing and eliciting an immune response, the specificity of which may be directed to the hapten carrier protein or new antigenic determinants in the conjugate, either singly or in combination (Bernstein,

Table 3.2 Classification of mechanisms of occupational asthma*

Non-immunologic mechanisms
    Reflex bronchoconstriction
    Irritant bronchoconstriction (i.e. reactive airways dysfunction syndrome)
    Pharmacologically induced bronchoconstriction
Immunologic mechanisms
    Immediate hypersensitivity response (IgE-mediated)
    High-molecular-weight allergens
    Low-molecular-weight allergens
        Mixed (immunologic and non-immunologic) allergens
        IgG antibody and/or immune (antigen–antibody) complex response
        Complement pathway activation
        Cell-mediated response

* Borrowed with permission from: Bernstein, J. A. (1992) Occupational asthma. *Postgraduate Medicine*, **92**, 109–118.

Table 3.3 Causes of occupational asthma: allergic or possibly allergic mechanism, low-molecular-weight compounds*

Agent	Occupation	# sub.	Prev. (%)	ST	IgE	ppt.Ab	Bronchoprovocation
Diisocyanates							
Toluene	Polyurethane industries, plastics, vanish	4	38.0				+ (4/4)
		21		−		−	
		112	12.5	+		−	+ (5/11)
		23	17.4	+	+		
		15			−		
		26			+		+ (26/26)
		17			−		+ (14/17)
		195	28.0		+		+ (12/17)
Diphenylmethane diisocyanate	Foundries	57	5.0	−	+		+ (6/11)
		1		−	+		+
		11					
Hexamethylene diisocyanate	Automobile spray painting	1			−	+	
Anhydrides							
Phthalic anhydride	Epoxy resins, plastics	4					+ (3/3)
		1		+	+		+
Trimellitic anhydride	Epoxy resins, plastics	14	29.0	+	+		+ (1/1)
Tetrachlorophthalic anhydride	Epoxy resins, plastics	14	36.0		−	−	+ (3/3)
		5					
Wood dust							
Western red cedar (*Thuja plicata*)	Carpenter, construction, cabinet maker, saw-mill worker	6					
		1320	3.4	+			+ (18/22)
		22		+	+	−	+ (185/185)
		185					
California redwood (*Sequoia sempervirens*)		2		−		−	+ (2/2)
Cedar of Lebanon (*Cedra libani*)		6		−		−	

Cocabolla (*Dalbergia resusa*)		2		–	+				+
Irolo (*Chlorophora excelsa*)		1		+	–		+	+	+
Oak (*Quercus robur*)		1		–	+		+	+	+
Mahogany (*Shorrea* sp.)		1		–	–		+	–	+ (2/2)
Abiruana (*Pouteria*)		2		+	+	+	–	+	+ (2/2)
African maple (*Triplochiton scleroxylon*)		2		+	+	–	–	+	+ (2/2)
Tanganyika aningre		3		+	+	+	–	+	+ (3/3)
Central American Walnut (*Juglans olanchana*)		1		–	+	–			+
Kejaat (*Pterocarpus angolensis*)		1		+	+	+			+
African zebra wood (*Microberlinia*)		1		+	+	+	–		+
Metals									
Platinum	Platinum refinery	91	57.0	+					+
		16		+				+	+ (10/16)
Nickel	Metal plating	1		+				+	+
		1		+			–	+	+
		1		+			–	+	+
Chromium	Tanning	1		–		–	+		+
		1		+		+	+		
		1		+		+			
Cobalt	Hard metal industry	4		+				+	
Vanadium		12	33.0	+			+		+ (2/4)
Tungsten carbide		1							
Fluxes									
Aminoethyl ethanolamine	Aluminium soldering	3							+ (3/3)
Colophony	Electronic	2		–					+ (2/2)
		51							+ (34/51)
Drugs									
Penicillins	Pharmaceutical	4		–					+ (3/4)
Cephalosporins	Pharmaceutical	2		+					+ (2/2)

(*Continued*)

Table 3.3 (Continued)

Agent	Occupation	#sub.	Prev. (%)	ST	IgE	ppt.Ab	Bronchoprovocation
Phenylglycine acid chloride	Pharmaceutical	24	29	+	+		+ (2/2)
Piperazine hydrochloride	Chemist	2		+			+ (2/2)
Psyllium	Laxative manufacturer	3		+			+ (2/3)
Methyldopa	Pharmaceutical	1		−			+
Spiramycin	Pharmaceutical	1		+			+
Salbutamol intermediate	Pharmaceutical	1					+
Amprolium HCl	Poultry feed mixer	1					+
	Pharmaceutical						+
Tetracycline	Manufacturer	1		+			
Sulphonechloramides	Brewery	12		+			
		7					
Other chemicals							
Dimethyl ethanolamine	Spray painting	1					+
Persulphate salts and henna	Hairdressing	2		−			+ (2/2)
Ethylene diamine	Photography	1		+			+
Azodicarbonamide	Plastics and rubber	151	18.5	−			
Dioazonium salt	Photocopying and dye	1					+
Hexachlorophene (sterilizing agent)	Hospital staff	1					+
Formalin	Hospital staff	28	29	−			+ (2/4)
Urea formaldehyde	Insulation, resin	2					+ (2/2)
Freon	Refrigeration	1					+
		1					+
Paraphenylene diamine	Fur dyes	80	37.5	+			+
Furfuryl alcohol (furan-based resin)	Foundry mould makers	1					+

* Adapted from Chan-Yeung M. et al. (1986) *American Review of Respiratory Disease*, **137**, 686.
# sub. = number of subjects; Prev. = prevalence; ST = skin testing; ppt.Ab = precipitating antibodies

IL, 1994; Bernstein, JA and Bernstein, 1994). Table 3.3 summarizes many of the common LMW agents causing immunologically mediated OA (Chan-Yeung and Lam, 1986).

Low-molecular-weight agent-induced OA has been associated with IgE-mediated, cytotoxic-mediated, immune complex-mediated and cell-mediated immune responses. IgE-mediated immune responses have been demonstrated for acid anhydrides, polyisocyanates and several other reactive chemical agents (Table 3.2) (Gallagher et al., 1981; Bernstein, IL, 1982; Bernstein, DI, et al., 1984). Cytotoxic immune responses, which involve the activation of classical complement pathways by IgM or IgG antibodies adherent to specific tissue antigens, have been postulated to cause a Goodpasture-like syndrome in workers exposed to high concentrations of trimellitic anhydride (TMA) (Zeiss and Patterson, 1993). These workers present with anaemia and pulmonary haemorrhage. Immune complex-mediated reactions have been postulated to be involved in the pathogenesis of chemical- and drug-induced pulmonary lesions, presumably as a result of antibody–antigen complex deposition in the lungs and consequent activation of complement pathways. The late respiratory systemic syndrome (LRSS) characterized by cough, wheezing, dyspnoea, mucus production and systemic symptoms of malaise, chills, myalgias and arthralgias occurring 4 to 12 hours after TMA inhalation exposure is believed to occur through this mechanism although cell-mediated mechanisms may also be involved (Zeiss and Patterson, 1993).

Cell-mediated responses have been increasingly recognized as important in the pathogenesis of OA induced by several LMW agents such as toluene diisocyanate (TDI) and plicatic acid. Low-molecular-weight chemicals form immunogenic conjugates which are then processed by antigen-presenting cells (APC) (Bernstein, JA and Bernstein, 1994). Relevant peptide determinants expressed on the APC cell surface in conjunction with class I or class II MHC antigens interact with T lymphocytes resulting in activation and release of lymphokines which set into motion the inflammatory cascade of events characteristic of asthma (Bernstein, JA and Bernstein, 1994; Bernstein, JA et al., 1996). Evidence for T-lymphocyte involvement in OA was first shown by Avery et al. who demonstrated lymphoblast transformation and lymphocyte proliferation in workers with TDI-OA (Avery et al., 1969). Gallagher et al. (1981) confirmed this observation by demonstrating the presence of lymphocyte-derived leucocyte inhibitory factor activity in the majority of their TDI-OA subjects. Subsequently, an increase in circulating peripheral and bronchial alveolar lavage (BAL) $CD8^+$ T cells and eosinophils was demonstrated in workers with TDI-induced OA (Finotto et al., 1991). T-lymphocyte involvement has also been shown in OA induced by nickel and cobalt metallic salts (Kusaka et al., 1989; Herzog et al., 1991). More recently, it has been demonstrated that workers with occupational exposure to opiates have changes in their peripheral lymphocyte populations. The significance of this observation to the underlying pathogenesis of opiate-induced OA is still unclear (Biagini et al., 1995).

Berylliosis is a chronic interstitial pneumonitis which occurs in workers exposed to beryllium who are employed in the manufacturing of alloys, ceramics, high technology electronics and fluorescent lighting and acts through cell-mediated immune mechanisms as well. The immunologic findings are very similar to those found for hypersensitivity pneumonitis (HP). Beryllium acts as a MHC class II restricted antigen to stimulate proliferation of $CD4^+$ Th1-like or delayed hypersensitivity cells localized in the lungs. These $CD4^+$ cells are primed to recognize beryllium and release interleukin 2 (IL-2), which leads to cellular proliferation, formation of granulomas and lung fibrosis (Kriebel et al., 1988; Rossman et al., 1988; Saltini et al., 1989; Bernstein, IL, 1994).

The observation that T-cell immune responses may be involved in the pathogenesis of occupational respiratory diseases is supported by animal models (Dearman et al., 1992). Rodents can be easily sensitized to LMW chemicals such as acid anhydrides and diisocyanates. A murine model has been used to study the differential production of cytokines by helper T-cell subsets (Th1 and Th2) after exposure to different reactive chemicals. It has been shown that under normal conditions Th1 lymphocytes primarily secrete IL-2, interferon-γ and tumour necrosis factor β (TNF-β) whereas Th2 lymphocytes secrete IL-4, IL-5, IL-10 and IL-13. A predominance of Th2 cells correlated with IgE-mediated immune responses, while a predominance of Th1 cells was associated with cell-mediated immune responses (Dearman et al., 1992). These investigators demonstrated preferential activation of Th2 cells by phthalic anhydride (PA), trimellitic anhydride (TMA) and diphenyl-methylene diisocyanate, indicating that chemical structure may be important for determining the quality of immune response elicited (Dearman et al., 1992).

Non-immunologic OA induced by LMW agents may mimic immunologically mediated OA. A possible non-immunologic mechanism may involve direct injury to bronchial epithelial cells by LMW agents such as acid anhydrides, diisocyanates, plicatic acid (the organic acid responsible for red cedar OA) or by environmental irritants such as ozone and sulphur dioxide (Chan-Yeung et al., 1973; Bernstein, IL, 1982; Brooks and Bernstein, 1993). The ensuing widening of gap junctions between bronchial epithelial cells permits easier access of these agents into the underlying lamina propria where they exert direct effects on inflammatory cells and the bronchial microvasculature and potentiate allergic sensitization (Brooks and Bernstein, 1993).

Another postulated non-immunologic mechanism responsible for LMW-induced OA may involve activation of the alternate complement pathway with subsequent increased production of anaphylatoxins (C3a, C5a), which then could stimulate release of bioactive mediators from mast cells. This possible mechanism has been reported in a few workers with red cedar OA but has not been substantiated for other causative agents of OA (Chan-Yeung, 1982).

A 'pharmacologic' non-immunologic mechanism has been ascribed to organophosphate insecticides which have anticholinesterase activity resulting in hypercholinergic-induced NSBH (Bernstein, DI, 1992). An outbreak of asthma was reported in workers involved in the manufacture of opioid drugs, which are known to be potent histamine-releasing agents (Biagini et al., 1990). A pharmacologic mechanism may partially explain the asthma-like symptoms of byssinosis after inhalation of cotton dust byproducts containing bacterial endotoxin (McKerrow et al., 1958; Bake et al., 1987; Merchant and Bernstein, 1993). A pharmacologic mechanism has also been proposed for meat wrapper's asthma (Vandervort and Brooks, 1977).

Non-specific bronchial hyperresponsiveness is a consequence of both immune and non-immune mediated OA. It may occur as a result of direct stimulation of sensory afferent nerves or by release of cytokines, neuropeptides and bioactive mediators from bronchial epithelial and/or inflammatory cells. These mediators have a host of chemotactic and inflammatory properties. Support for a neurophysiologic mechanism of OA was demonstrated in a guinea pig model (Thompson et al., 1987; Sheppard et al., 1988). Exposure of guinea pig afferent nerve endings to TDI resulted in the release of tachykinins (substance P, neurokinin A) and development of NSBH. The NSBH was potentiated by TDI's ability to inhibit neutral endopeptidase, an ectoenzyme important for tachykinin degradation (Sheppard et al., 1988). Phthalic anhydride (PA) also has been shown to exhibit a neurogenic effect by stimulating local release of substance P and calcitonin gene-related peptide (CGRP) from afferent nerve endings independent of the central nervous system (Thompson et al., 1987).

3.5 Pathophysiology

Chemically-induced respiratory diseases appear to have affinities for specific anatomical compartments of the lungs. Therefore, it is reasonable to discuss the pathophysiology of these disorders in the context of their anatomical localization.

3.5.1 *The Conducting Airways*

Upper airways

The nose and nasopharynx are gradually becoming recognized as important target organs for chemically-induced hypersensitivity responses. It has been well established that allergic rhinitis is a precursor of asthma in work-related respiratory syndromes caused by HMW proteins including laboratory animal-induced asthma and bakers' asthma. However, rhinoconjunctivitis symptoms are also commonly seen in chemically induced respiratory syndromes. Acid anhydride and platinum refinery workers have a high incidence of nasal symptoms with or without asthma (Bernstein, IL and Brooks, 1993; Zeiss and Patterson, 1993). In many instances, positive IgE-mediated skin prick tests to these agents can be demonstrated. A recent prospective study of diisocyanate-exposed workers revealed a surprisingly high incidence of nasal symptoms irrespective of whether or not they had asthma (Bernstein, DI, 1993). A similar finding was found in a group of petrochemical workers who were assessed for non-specific nasal responsiveness (Plavec et al., 1993). These investigators used nasal provocation testing to demonstrate that these workers exhibited a greater degree of nasal hyperresponsiveness during periods of continuous exposure to non-specific irritants followed by resolution of symptoms when they were removed from further exposure to these irritants (Plavec et al., 1993). Nasal provocation testing and nasal lavage have been useful tools for evaluating the pathophysiology of nasal reactions (Plavec et al., 1993; Ahman et al., 1995).

Lower airways

As discussed earlier OA is the most common occupational respiratory disorder. Specific bronchoprovocation studies, bronchoalveolar lavage (BAL) and transbronchial biopsy (TBBx) have served as useful techniques for studying the pathophysiology of respiratory disorders induced by chemicals. Specific inhalation challenge tests have been demonstrated to correlate well with the worker's occupational history and the histopathologic changes found by BAL and TBBx (De Monchy et al., 1985; Chan-Yeung et al., 1988). Workers sensitized to LMW chemical agents may exhibit diverse patterns of bronchospastic reactions. Such workers manifest an early asthmatic reaction (EAR) 10 to 20 per cent of the time, an isolated late asthmatic reaction (LAR) 30 to 50 per cent of the time and a dual asthmatic reaction (DAR) 30 to 50 per cent of the time (Fabbri et al., 1993). Late asthmatic reactions occur 2 to 12 hours after the early asthmatic response subsides and are associated with the influx of inflammatory cells into the airways (Mapp et al., 1986; Fabbri et al., 1993). However, recent studies have demonstrated that an increase in NSBH may actually precede cellular infiltrates by as early as 2 hours post-antigen challenge (Durham et al., 1987; Thorpe et al., 1987). Figure 3.1 illustrates a characteristic DAR bronchoprovocation response in a worker with TDI-OA (Bernstein, DI, 1992). It is now well recognized that asthma may persist for years even after removal from further exposure. The prolonged morbidity which may result from OA has been found to correlate directly with the length of time the worker is exposed to the causative chemical agent prior to diagnosis and removal from further exposure (Chan-Yeung, 1993).

Histologic features in the airways of patients with OA are similar to those found in individuals with non-occupational asthma (NOA). Common pathologic sequelae include airway smooth muscle hypertrophy, mucosal oedema, increased mucus production from greater numbers of mucous secreting glands and goblet cells, and deposition of collagen beneath the basement membrane (Fabbri et al., 1993). BAL fluids from normal, NOA and OA subjects have similar inflammatory-cell count distributions (90 to 95 per cent alveolar

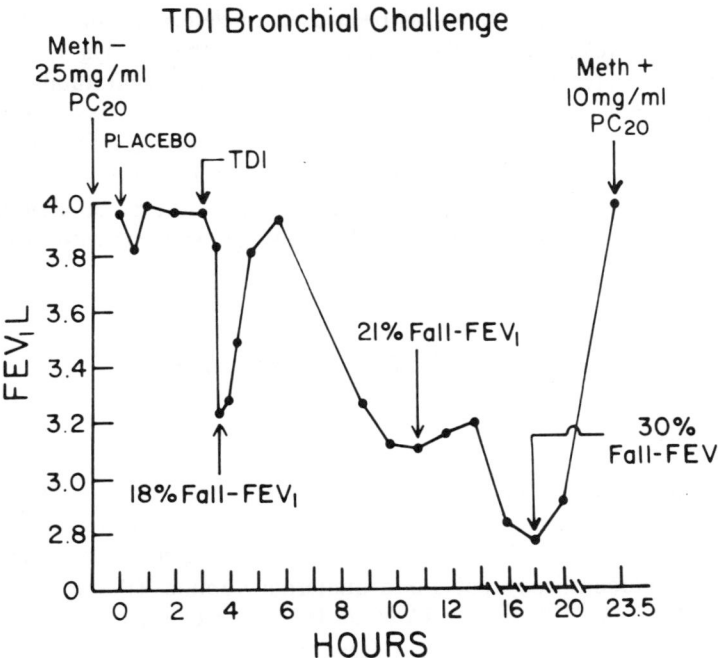

Figure 3.1 A dual-phase response in a polyurethane foam worker after bronchial provocation with toluene diisocyanate (TDI). Early-phase bronchospasm is followed by spontaneous recovery, which in turn is followed by late-phase bronchospasm. The concentration of methacholine required to reduce the forced expiratory volume in 1 second (FEV_1) by 20 per cent (PC_{20}) fell from a baseline normal of 25 mg . ml^{-1} to 10 mg . ml^{-1} the day after TDI challenge (with permission from Bernstein, JA et al., 1996).

macrophages, 5 to 10 per cent lymphocytes and less than 1 per cent leucocytes and epithelial cells) (Fabbri et al., 1993). However, these cell numbers dramatically increase in those individuals with persistent bronchial hyperreactivity or active OA. Saetta *et al.* examined serial bronchial biopsies in workers with TDI-OA and found increased numbers of inflammatory cells compared with normal non-asthmatic subjects (Saetta et al., 1992a). The lamina reticularis beneath the basement membrane was thickened due to deposition of types 1, 3 and 5 collagen (Saetta et al., 1992a, 1992b). It has also been observed that junctional gaps between basal epithelial cells are widened (Wegner et al., 1984). Such defects would presumably contribute to the loss of intercellular adhesion molecules important for regulating migration of inflammatory cells into the airways. TBBx from patients with OA have demonstrated increased numbers of eosinophils and activated T cells (Saetta et al., 1992a, 1992b). *In situ* hybridization studies also revealed expression of mRNA for IL-5, the cytokine responsible for growth, differentiation and activation of eosinophils (Saetta *et al.*, 1992a).

Bronchoalveolar lavage has also been used to detect the presence of bioactive mediators released by inflammatory cells into the airways. Leukotrienes (LTE_4) were detected immediately after TDI and plicatic acid inhalation challenges but disappeared during the LAR (Chan-Yeung et al., 1989; Zocca et al., 1990). Microvascular leakage, as measured by increased amounts of albumin in BAL, is also a useful marker for active asthma. Other preformed (histamine) and newly formed mediators (platelet activating factor, prostaglandins and leukotrienes) have been found in BAL obtained from subjects with OA (Chan-Yeung et al., 1989).

3.5.2 Alveolar, Interstitial and Capillary Compartments

Alveolitis

Extrinsic allergic alveolitis or hypersensitivity pneumonitis (HP) was first described in farmers after exposure to thermophilic actinomycetes (Pepys et al., 1969). Most cases of HP occur after exposure to organic dust but causative agents also include diisocyanates and quaternary amines (Blake et al., 1965; Malo and Zeiss, 1982; Bernstein, JA et al., 1994). The characteristic pathologic lesions associated with HP include diffuse inflammation and pneumonitis of the interstitium, alveoli and terminal bronchioles (Bernstein, IL, 1994). Chemically-induced HP occurs more acutely or subacutely after exposure compared with HP induced by organic dusts where farmers present with chronic lung changes (Bernstein, IL, 1994). Chest radiographs may be normal or reveal fine nodular densities or large peripheral lung infiltrates (Bernstein, IL, 1994). Bronchoalveolar lavage (BAL) reveals increased numbers of $CD8^+$ activated T-lymphocytes (Bernstein, IL, 1994). Precipitating antibodies have been found in HP induced by diphenyl-methylene diisocyanate (Zeiss et al., 1980).

Alveolar haemorrhage

Pulmonary haemorrhage associated with haemolytic anaemia has been reported in four men exposed to very high concentrations of heated trimellitic anhydride fumes (Zeiss and Patterson, 1993). Increased total antibody levels were found in these men. Using a rat model it has been shown that there was a linear relationship between antibody concentrations and the degree of pulmonary haemorrhage suggesting that immune mechanisms may play a role in this disorder (Zeiss and Patterson, 1993).

Granulomatosis and fibrosis

Acute berylliosis occurs after an unusually large exposure to beryllium and may lead to chemical pneumonitis. After exposure subsides, lung lesions disappear and there are few permanent residual effects (Kriebel et al., 1988). However, chronic exposure to beryllium leads to lung disease associated with an immune pathogenesis (Kriebel et al., 1988). Symptomatic workers demonstrate delayed hypersensitivity reactions in skin tests and *in vitro* specific lymphocyte proliferation (Rossman et al., 1988; Saltini et al., 1989; Bernstein, IL, 1994). Bronchoalveolar lavage studies reveal large increases in activated MHC class II restricted $CD4^+$ cell numbers (Rossman et al., 1988; Saltini et al., 1989; Bernstein, IL, 1994). Characteristic lung changes reveal diffuse interstitial infiltration of mononuclear cells with or without non-caeseating granulomata. Lung nodularity is often observed and as the disease progresses interstitial fibrosis occurs, resulting in a reduction of diffusing capacity (Bernstein, IL, 1994).

There are several clinical reports suggesting an increased prevalence of pulmonary fibrosis and scleroderma among underground gold miners and other workers heavily exposed to silicosis (Erasmus, 1957). The immunopathogenesis of these associations remains to be elucidated.

3.6 Specific Causes of Chemical-Induced Respiratory Disease

3.6.1 Polyisocyanates

It is estimated that 5 to 10 per cent of workers exposed to diisocyanates will develop OA. The reactive $N=C=O$ side groups of these compounds make them useful in the production of polyurethane foams, plasticizers, adhesives, coatings and paint hardeners used in automotive, building and spray paint industries (Bernstein, JA et al., 1996). The three major commonly used diisocyanate compounds in industry are hexamethylene diisocyanate (HDI), 4-4-diphenylmethylene diisocyanate (MDI) and toluene diisocyanate (TDI) (Bernstein, JA et al., 1996). Most reports of diisocyanate OA have resulted after a large exposure from an accidental spill (Mapp et al., 1988). Only 10 to 15 per cent of diisocyanate workers with OA will show evidence of IgE sensitization by *in vitro* tests (Bernstein, JA et al., 1996). Even though the chemical structures of diisocyanate compounds vary, sensitized workers who produce antibody often exhibit significant cross-reactivity to other chemicals in this class as demonstrated by RAST inhibition studies (Baur, 1983; Kennedy and Brown, 1992). Workers with diisocyanate-induced OA and specific antibodies frequently manifest heterogeneous isotypic antibody responses which appear to depend on their exposures to specific diisocyanates (Baur, 1983; Moller et al., 1986a; Kennedy and Brown, 1992). For example, HDI-exposed subjects primarily produce specific IgE and IgG antibodies compared with MDI workers who produce IgE, IgG and/or IgM, and TDI workers who frequently do not have a detectable antibody response (Bernstein, JA et al., 1996). Serological assays measuring antibody specificity to diisocyanate conjugated-albumin reagents are more sensitive than skin testing and are therefore more reliable as a screening test for detecting specific antibodies in these workers. This is the exact opposite of antibody responses in workers sensitized to HMW molecules. Diisocyanates are also known to cause RADS, hypersensitivity pneumonitis and, rarely, bronchiolitis obliterans (Gallagher et al., 1981; Luo et al., 1990; Tarlo and Broder, 1989).

3.6.2 Acid Anhydrides

Acid anhydrides are a group of reactive organic compounds which are constituents of alkyl and epoxy resins. These chemicals are found in paints, varnishes and plastics. TMA has been very useful as a plasticizer for the production of wire and cable coatings and has wide applications in the manufacturing of other products such as paints, textiles and paper goods (Zeiss and Patterson, 1993). Workers exposed to acid anhydrides frequently produce IgE antibodies to either the ligand (acid anhydrides) or new antigenic determinants created after their conjugation to a protein carrier, and, therefore, in vitro assays such as the radioallergosorbent test (RAST) and enzyme-linked immunosorbent assay (ELISA) in conjunction with skin testing to acid anhydride conjugates have served as sensitive markers for disease (Zeiss et al., 1980; Zeiss and Patterson, 1993). Increased IgG levels have also been reported and in one study elevated IgG4 levels suggested the possibility that this isotype could be involved in the underlying immunopathogenesis of OA (Zeiss and Patterson, 1993). As cited earlier, some workers develop a late respiratory systemic syndrome that has been associated with IgG and IgA antibodies, or the pulmonary haemorrhage/anaemia syndrome that has been associated with elevated total antibody levels (Patterson et al., 1982).

3.6.3 Wood dusts

Wood dusts are responsible for several occupationally-related lung disorders including hypersensitivity pneumonitis, organic dust syndrome, asthma, chronic bronchitis and a mucous membrane irritation syndrome (Chan-Yeung et al., 1973; Enarson and Chan-Yeung, 1990). Western red cedar (*Thuja plicata*) has been most extensively studied. Prospective studies of workers exposed to western red cedar dust reported that approximately 5 per cent develop OA (Chan-Yeung et al., 1987). Western red cedar wood dust contains a variety of chemicals including tannin, dyes, pitch, resins and gums. These chemicals are extracted by steam distillation from the cedar wood to form volatile and non-volatile components. The volatile components contain a variety of agents such as tropolones and natural fungicides which protect the wood from decay. Non-volatile residues consist primarily of plicatic acid which is the organic component linked to causing most cases of red cedar-induced OA (Bernstein, JA et al., 1996). Specific inhalation tests to plicatic acid

Occupational Respiratory Allergy

typically reveal dual asthmatic reactions (Chan-Yeung et al., 1987). Since immediate skin testing and serum IgE antibodies are not diagnostically useful, specific bronchial challenge using very low concentrations of plicatic acid is often necessary to make a definitive diagnosis (Paggiaro and Chan-Yeung, 1987). Several other wood dusts have also been recognized as causing OA, including African maplewood and zebrawood, mahogany and Quillaja bark (Chan-Yeung, 1993). Wood dusts are also responsible for causing hypersensitivity pneumonitis disorders including maple bark disease, sequoiosis suberosis and wood pulp workers' disease (Chan-Yeung, 1993). Organic acids have not been implicated in these diseases, which are caused by sensitization to a variety of contaminating fungi including *Cryptostroma corticale* (maple bark), *Alternaria*, *Aspergillus* and the bacterium, *Thermoactinomyces vulgaris*. Wood dusts have also been suggested as causing other non-respiratory symptoms including allergic conjunctivitis, rhinitis, dermatitis and adenocarcinoma of the nasopharynx (Chan-Yeung, 1993).

3.6.4 Metallic Salts

Metallic salts cause a variety of pulmonary reactions including asthma, chemical pneumonitis, bronchitis, pulmonary emphysema and adult respiratory distress syndrome (ARDS) (Nemery, 1990). Metallic salts induce respiratory diseases through immunologic and non-immunologic mechanisms. Those metallic salts most extensively studied include platinum, nickel, chromium and cobalt salts (Bernstein, IL and Brooks, 1993). Platinum salts are highly allergenic compared with other metallic salts. Cigarette smoking has been implicated as a significant risk factor for platinum sensitization (Biagini et al., 1985). The immunological pathogenesis of platinum salt-induced OA has been demonstrated to be IgE-mediated but may also involve reaginic IgG4 antibody. This has been supported by skin testing, RAST inhibition and specific IgE radioimmunoassay studies (Biagini et al., 1985). The effect of platinum salt-induced bronchial hyperresponsiveness (BHR) may be potentiated by concomitant exposure to non-specific irritants, such as chlorine gas commonly found in platinum refineries (Biagini et al., 1985).

Nickel is widely used in mining, milling, smelting and refinishing processes and can be extremely toxic to the central nervous system, but nickel-induced OA is very rare. Nickel-specific IgE antibodies have been found in a few individual case reports (Bernstein, IL and Brooks, 1993).

Chromium is used in electroplating processes, leather tanning and in the production of cement and has been occasionally implicated as a cause of OA (Burrows, 1984). IgE-specific antibody and cell-mediated immune responses have both been postulated as underlying mechanisms in causing pulmonary disease (Burrows, 1984; Moller et al., 1986b).

Cobalt salts are responsible for inducing OA in approximately 5 per cent of exposed workers and should be distinguished from hard metal (mixture of cadmium and cobalt) diseases that result in interstitial pneumonitis and pulmonary fibrosis (Gheysens et al., 1985; Shirakawa et al., 1989). Workers who are exposed to cobalt salts alone and subsequently develop OA have been documented to produce specific IgE antibodies (Shirakawa et al., 1989).

A variety of other metals has been associated with respiratory diseases, including vanadium pentoxide, zinc (metal fume fever), cadmium and aluminium (potroom asthma) (Bernstein, IL and Brooks, 1993). Individuals with metal fume fever and potroom asthma both may present with shortness of breath, wheezing, chest tightness and/or coughing that occur hours after exposure. Symptoms of metal fume fever usually subside 1 to 2 days after removal from the workplace, while the clinical course of potroom asthma is more unpredictable as symptoms may persist long after removal from exposure (Bernstein, IL and Brooks, 1993). Since the clinical spectrum of this group of metal-induced diseases is so variable and immunologic responses have not been found, they should be referred to as either non-immunologic asthma or asthma-like syndromes.

3.6.5 Colophony

Colophony fumes may cause OA in electronic soldering workers (Burge, 1982a, 1982b). Colophony is derived from pine tree resin and contains abietic, pimaric and dihydroabietic acids which are believed to be responsible for inducing OA (Burge, 1982a, 1982b). Atopic workers with pre-existing asthma are at an increased risk for developing OA from colophony fumes (Burge, 1982a, 1982b). Mechanisms for colophony-induced OA are unknown and diagnosis often depends on specific inhalational challenges. The persistence of NSBH is variable among workers once removed from the workplace (Burge, 1982a, 1982b).

3.6.6 Sulphonechloramide

Sulphonechloramide (chloramine-T), an antibacterial used in the food and brewing industry, has been shown to induce IgE-mediated OA which can be detected by skin testing using this agent alone without conjugation to a protein carrier (Malo and Bernstein, 1993).

3.6.7 Miscellaneous Agents

A number of other LMW chemicals have been associated with occupational respiratory hypersensitivity disorders including asthma. For the most part these are anecdotal case reports of work-associated symptoms developing after exposure to primary, secondary, tertiary and quaternary amines (Malo and

Bernstein, 1993; Bernstein, JA *et al.*, 1994), formaldehyde (Malo and Bernstein, 1993), glutaraldehyde (Burge, 1989), persulphates (Pepys *et al.*, 1976; Baur *et al.*, 1979), diazonium salts (Luczynska *et al.*, 1990), reactive azo and anthraquinone dyes (Alanko *et al.*, 1978; Docker *et al.*, 1987) and a variety of pharmaceutical agents (i.e. β-lactam, macrolide and tetracycline antibiotics, hydralazine, isoniazid, penicillamine and methyldopa) (Cooper *et al.*, 1986a, 1986b). In addition, pyrolysis fumes of polyvinyl chloride (PVC) have been linked with the asthma-like syndromes of meat wrapper's asthma (Brooks and Vandervort, 1977; Vandervort and Brooks, 1977; Pauli *et al.*, 1980). Other chemicals which have been associated with causing asthma include acrylates, freon, hexachlorophene, furan, styrene and machining fluids (Malo and Bernstein, 1993; Bernstein, DI *et al.*, 1995). The mechanisms underlying these responses are unclear and at least in the case of the latter two agents, contaminants during the manufacturing processes may play a role (Bernstein, DI *et al.*, 1995).

3.7 Miscellaneous Occupational Respiratory Disorders

3.7.1 *Reactive Airways Disease (RADS)*

Discussion of chemically-induced respiratory disorders would be incomplete without reference to the reactive airways dysfunction syndrome, also referred to as irritant-induced occupational asthma. Reactive airways dysfunction syndrome occurs after a single exposure to high levels or after chronic low-level exposure to irritant vapours, fumes, gas or smoke and leads to persistent NSBH (Brooks *et al.*, 1985; Tarlo and Broder, 1989; Brooks, 1994). It is characterized by the absence of a latency period but is histopathologically similar to asthma and other forms of OA (Brooks and Bernstein, 1993). One striking feature of RADS which may differentiate this disorder from other forms of OA and non-occupational asthma is that the depth of collagen deposition beneath the reticular basement membrane tends to be greater than that observed for other forms of asthma (Gautrin *et al.*, 1994). The mechanism(s) and natural history of RADS are still not well defined but it is well established that NSBH may persist in affected individuals for months to years after their initial insult (Brooks and Bernstein, 1993).

3.7.2 *Multiple Chemical Sensitivity Syndrome (MCS)*

The multiple chemical sensitivity (MCS) syndrome is a term that has been used to describe a constellation of non-specific systemic symptoms occurring after exposure to a wide variety of irritating agents. Patients present with headaches, food intolerance, fatigue, lack of concentration and depression which they attribute to their work environments. These symptoms subside

Table 3.4 Case definition of multiple chemical sensitivity*

(1) Initial symptoms acquired in relation to an identifiable environmental exposure(s)
(2) Symptoms involve more than one organ system
(3) Symptoms recur and abate in response to predictable stimuli
(4) Symptoms are elicited by low-level exposures to chemicals of diverse structural classes
(5) No standard test of organ system function can explain symptoms

* Adapted from Fiedler, N., Maccia, C. and Kipen, H. (1992) Evaluation of chemically sensitive patients. *Journal of Occupational Medicine*, **34**, 529–538.

away from work and reappear after returning to the work site (Lax and Henneberger, 1995). The heterogeneous spectrum of these symptoms has made it very difficult to design scientifically valid epidemiological investigations of these individuals. Immunologic evaluation is usually normal (Simon *et al.*, 1993). Intensive searches for environmental factors at work and home that aggravate these symptoms have been unsuccessful in identifying specific causes. In 50 per cent of suspected cases of MCS, some form of psychoaffective disorder has been identified by psychometric analysis (Simon *et al.*, 1993). A suggested case definition of MCS syndrome is summarized in Table 3.4 (Fiedler *et al.*, 1992).

3.7.3 Sick Building Syndrome (SBS)

Sick building syndrome is the term used to describe symptoms and health problems which workers attribute to their work environment and poor indoor air quality. According to a World Health Organization working group, this syndrome is characterized by eye, nose and throat irritation, sensation of dry mucous membranes and skin, erythema, mental fatigue, headache, increased frequency of airway infections and cough, hoarseness, wheezing, itch, nausea and dizziness (Seltzer, 1994). Room temperatures greater than 22°C and a relative humidity of less than 25 per cent have been associated with symptoms of SBS in patients who become ill working in a 'sick' building (Jaakkola *et al.*, 1994). These symptoms were alleviated with modest air humidification. Non-hygienic air-conditioning and humidification systems, poorly maintained ducts and plumbing leaks can be a source of microbiological contamination. Overcrowding of work space by personnel and/or clients may cause increased levels of carbon dioxide. Other potential causes of symptoms include volatile hydrocarbons, photocopiers and related materials, textiles, soft fibre wall materials and organic gases and particulates found in tobacco smoke (Jaakkola *et al.*, 1994; Seltzer, 1994).

Several investigations of volatile organic compounds (VOC) have been reported. Controlled human chamber exposures to a VOC mixture ($25 \text{ mg} \cdot \text{m}^{-3}$) in healthy volunteers revealed that symptom and irritation intensity was unrelated to detection of aversive odour and more likely related to stimulation of trigeminal nerve receptors (Hudnell *et al.*, 1992). Moreover, similar VOC exposures evoked significant increases of neutrophils in nasal lavages obtained after a 4-hour exposure (Koren et al., 1992). In analyzing most published reports of SBS, women seem to report symptoms more often than men. Psychosocial factors such as job satisfaction, work stress and social atmosphere may also be major determinants of symptoms (Seltzer, 1994).

3.8 Diagnosis of Chemical Respiratory Disorders

The diagnosis of chemically induced respiratory disorders requires a careful occupational history (Bernstein, DI, 1993). A good history should try to establish an occupational relationship with the initiation of symptoms. For example, OA typically improves away from work and recurs upon re-exposure. However, there are several instances where a history may fail to identify the relationship between respiratory symptoms and work (Bernstein, DI, 1993). For example, workers suspected of having OA may manifest primarily a LAR and therefore may not have symptoms until they are at home or away from work. Also if OA is very severe there may not be noticeable improvement of symptoms away from work (Bernstein, DI, 1993).

Material safety data sheets (MSDS) must be accessible to employees in their work areas by federal law and are frequently useful for determining the most likely agent(s) responsible for inducing respiratory disease (Bernstein, JA, 1992). Industrial hygienists, who routinely perform air sampling measurements in the workplace in compliance with established standards (e.g. National Institute for Occupational Safety and Health (NIOSH)), are often helpful in providing information about exposure levels of specific agents (Bernstein, JA, 1992).

The diagnosis of OA requires objective evidence either by demonstrating the presence of significant reversibility of the FEV_1 (forced expiratory volume) by bronchodilators or the presence of NSBH by methacholine or histamine challenge tests (Bernstein, DI, 1993). Occupational asthma must be distinguished from other lung diseases such as chronic obstructive lung disease, pneumoconiosis, bronchiolitis obliterans, hypersensitivity pneumonitis, organic dust syndromes and endotoxin-induced pulmonary syndromes that occur in cotton or swine herd confinement workers (Bernstein, DI, 1993).

Objective assessment of lung function in the workplace is essential for making a diagnosis of OA. It is important that assessment of lung function be made in the work environment as measurements before and after a workshift may not adequately establish a relationship between symptoms and work exposure (Bernstein, JA, 1992; Bernstein, DI, 1993). Improvement of symptoms and lung function after removal from the workplace, combined with subsequent deterioration after reintroduction to the workplace, also support a diagnosis of OA. If pulmonary function tests are normal, PD_{20} measurements (provocative dose of methacholine or histamine required to cause a 20 per cent decrease in FEV_1) in and away from the workplace are useful in demonstrating changes in NSBH (Bernstein, JA, 1992; Bernstein, DI, 1993).

The most frequently used method to measure lung function objectively at and away from work is performance of serial peak expiratory flow rates (PEFR) over periods of 2 to 3 weeks (Bernstein, DI, 1993). The subject measures and records his/her PEFR every 1 to 2 hours while at work and at home and at the same times documents symptoms and medication usage. A diurnal variability of greater than 20 per cent at work with subsequent improvement or normalization at home is considered an abnormal response consistent with OA. Ideally, PEFRs should be monitored for 2 weeks at work and during a 1 to 2 week period away from work. One of the shortcomings of these measurements is that patients can falsify or inaccurately record their readings (Bernstein, DI, 1993).

The most definitive method for establishing a diagnosis of OA is the specific inhalation challenge test (Cartier and Malo, 1993). If a specific substance in the workplace is suspected of causing OA, a specific inhalation challenge should ideally be performed to confirm this relationship, provided that this can be done without risk to the worker (Cartier and Malo, 1993). A preliminary measurement of NSBH (PD_{20}) can be helpful in estimating a safe initial dose of the challenge agent. These tests are time consuming and potentially dangerous and therefore should be performed only at specialized centres by experienced personnel. Challenges should not be performed in workers with severe cardiac or pulmonary (FEV_1 less than 60 per cent) diseases (Cartier and Malo, 1993). Specific inhalation challenge tests may be recommended to document suspected index cases due to xenobiotic substances which have not been previously described as causes of OA or for evaluating a worker's eligibility for worker's compensation. Although positive specific challenge tests provide a definitive diagnosis of OA, negative results do not always exclude the diagnosis because workers removed from the workplace for a long period of time may lose sensitivity to the suspected agent. It is therefore important to perform a challenge test not long after the worker is removed from the workplace. Another potential problem with specific inhalation challenge tests is the difficulty of reproducing work exposure conditions in the laboratory. Furthermore, a number of technical factors must be controlled during these challenges, including temperature, atmospheric pressure and concentration of the challenge agent. Ideally, specialized centres that perform challenge tests routinely and according to a standard protocol would greatly facilitate diagnosis and research of OA (Cartier and Malo, 1993).

Although chest radiographs are usually normal in OA, they may be helpful in making a diagnosis of pulmonary fibrosis or HP. However, there are no characteristic chest radiograph findings diagnostic of HP and even in symptomatic patients there may be no abnormalities (McCloud, 1991; Sharma, 1991). In acute HP, diffuse patchy infiltrates or discrete nodular infiltrates are often observed. In chronic HP the chest radiograph may reveal diffuse reticulonodular infiltrates and honeycombing. Chest radiograph findings for pulmonary fibrosis reveal diffuse reticular markings predominantly in the lower lung zones. High resolution CT scans are sometimes helpful for delineating bronchiectasis and interstitial fibrosis in advanced disease (McCloud, 1991; Sharma, 1991). Lung diffusing capacity (DLCO) is usually decreased in HP and pulmonary fibrosis but is normal in OA. Arterial gas tensions may be normal or decreased in any of these disorders depending on the stage of disease (Sharma, 1991).

In evaluating a worker for OA, it is important to determine the presence of atopy by skin testing with common aeroallergens and other appropriate allergens. *In vitro* assays to measure specific IgE to specific chemical agents are available for the most reactive agents. In contrast to natural inhalant and protein occupational allergens, these tests for chemical agents are generally more reliable than *in vivo* skin testing (Bernstein, DI, 1993). However, it should be emphasized that the presence of either positive prick skin tests or specific IgE in the serum only indicates IgE-mediated sensitization and does not in itself confirm a clinical diagnosis of OA. Specific IgG antibodies, when present, may be used as markers of exposure to a particular chemical antigen (Bernstein, DI, 1993). *In vitro* tests for cell-mediated immunity substantiate the association of an immunologic response. In specific situations (e.g. chronic berylliosis) specific lymphocyte proliferative assays indicate that delayed hypersensitivity is the underlying mechanism of the disease process. These assays have been used as early screening tests to determine if beryllium-exposed workers have become sensitized (Rossman *et al.*, 1988; Saltini *et al.*, 1989).

3.9 Treatment

Once the diagnosis of an occupational respiratory disorder has been confirmed, treatment should be directed toward removing the worker from further exposure. Studies evaluating the clinical course of workers with OA after removal from exposure have found that the persistence of asthma frequently depended on the duration of symptoms prior to diagnosis (Bernstein, JA *et al.*, 1996). Individuals diagnosed early in the course of their disease had relatively normal lung function and a lesser degree of airway hyperresponsiveness which tended not to persist after removal from the workplace (Bernstein, JA *et al.*, 1996). Individuals who remained in the workplace with continuous exposure to the offending agent remained stable on medication approximately 50 per cent of the time, whereas the remainder experienced deterioration of lung function with a significant increase in symptoms and medication requirements. Use of respirators in the work environment generally does not reduce exposure or prevent clinical deterioration (Bernstein, JA *et al.*, 1996). Some studies have suggested that certain types of respirators, such as airstream helmets, may offer adequate protection for the worker from the offending agent. However, they are generally not considered to be better than absolute avoidance measures (Bernstein, JA *et al.*, 1996). Treatment of acute or chronic OA is similar to the conventional treatment of asthma which involves selected β_2-agonists, theophylline, cromolyn sodium, inhaled and/or systemic corticosteroids in various combinations depending on the severity of the worker's symptoms (Bernstein, JA et al., 1996). Systemic corticosteroids are the only effective agent for the treatment of the alveolar and interstitial pneumonitides.

3.10 Prevention

The prevention of chemical-induced respiratory disease depends on cooperation between management and labour in organizing comprehensive immunosurveillance programmes for detecting and monitoring workers at increased risk for exposure to known causative agents (Bernstein, JA et al., 1996). Employment prescreening of workers to exclude risk factors contraindicated for certain jobs (i.e. atopy in acid anhydride workers) should be performed when appropriate. Every attempt should be made to minimize a worker's exposure to potentially problematic agent(s) through the institution of strict handling procedures (Bernstein, DI, 1993; Bernstein, JA et al., 1996). Workers should be educated continually about the importance of adhering to these procedures with special attention to avoiding large chemical spills. For OA, it has already been demonstrated that implementation of comprehensive immunosurveillance programmes in the workplace significantly reduces the incidence of disease (Bernstein, DI, 1993).

3.11 Conclusions

Recognition of the importance of chemical-induced respiratory diseases has improved as industry continues to introduce a greater number of potentially toxic and sensitizing compounds into the work environment. The diagnosis of these respiratory disorders requires a thorough history, physical examination, chest radiograph, appropriate objective measurements of lung function and immunological testing when appropriate. Prospective epidemiologic studies in large populations of workers, utilizing validated standardized questionnaires to identify workers at risk for respiratory disease, would aid in determining more accurate prevalence statistics of occupational respiratory diseases, in comparison with cross-sectional studies which historically have underestimated the number of affected workers. Prospective evaluations would also facilitate identification of specific risk factors and enhance understanding of the natural course and immunopathogenesis of these disorders.

References

AHMAN, M., HOLMSTROM, M. & INGELMAN-SUNBERG, H. (1995) Inflammatory markers in nasal lavage fluid from industrial arts teachers. *American Journal of Industrial Medicine*, **28**, 541–550.

ALANKO, K., KESKINEN, H., BJORKSTEN, F. & OJANEN, S. (1978) Immediate-type hypersensitivity to reactive dyes. *Clinical Allergy*, **8**, 25–31.
AVERY, S. B., STETSON, D. M., PAN, P. M. & MATHEWS, K. P. (1969) Immunological investigation of individuals with toluene diisocyanate asthma. *Experimental Immunology*, **4**, 585–596.
BAKE, B., RYLANDER, R. & FISCHER, J. (1987) Airway hyperreactivity and bronchoconstriction after inhalation of cell-bound endotoxin, in JACOBS, R. R. & WAKELYN, P. J. (Eds) *Proceedings of the Eleventh Cotton Dust Research Conference*, pp. 12–14. Memphis, TN: National Cotton Council.
BAUR, X. (1983) Immunologic cross-reactivity between different albumin-bound isocyanates. *Journal of Clinical Immunology*, **71**, 197–205.
BAUR, X., FRUHMANN, G. & LIEBE, V. V. (1979) Occupational asthma and dermatitis after exposure to dusts of persulfate salts in two industrial workers. *Respiration*, **38**, 144–150.
BECKLAKE, M. R. (1993) Epidemiology: prevalence and determinants, in BERNSTEIN, I. L., CHAN-YEUNG, M., MALO, J.-L. & BERNSTEIN, D. I. (Eds) *Asthma in the Workplace*, pp. 29–59. New York: Marcel Dekker.
BECKLAKE, M. R. & LALLOO, U. (1990) The 'healthy' smoker effect: a phenomenon of health selection, *Respiration*, **57**, 137–144.
BERNSTEIN, D. I. (1992) Occupational asthma. *Clinical Allergy*, **76**, 917–934.
(1993) Clinical assessment and management of occupational asthma, in BERNSTEIN, I. L., CHAN-YEUNG, M., MALO, J.-L. & BERNSTEIN, D. I. (Eds) *Asthma in the Workplace*, pp. 103–123. New York: Marcel Dekker.
BERNSTEIN, D. I., GALLAGHER, J. S., D'SOUZA, L. & BERNSTEIN, I. L. (1984) Heterogeneity of specific-IgE responses in workers sensitized to acid anhydride compounds. *Journal of Allergy and Clinical Immunology*, **74**, 794–801.
BERNSTEIN, D. I., LUMMUS, Z. L., SANTILLI, G., SISKOSKY, J. & BERNSTEIN, I. L. (1995) Machine operator's lung. *Chest*, **108**, 636–641.
BERNSTEIN, I. L. (1982) Isocyanate-induced pulmonary disease: a current perspective. *Journal of Allergy and Clinical Immunology*, **70**, 24–32.
(1994) Respiratory allergy: clinical aspects. *Toxicology in Vitro*, **8**, 975–980.
BERNSTEIN, I. L. & BROOKS, S. M. (1993) Metals, in BERNSTEIN, I. L., CHAN-YEUNG, M., MALO, J. L. & BERNSTEIN, D. I. (Eds) *Asthma in the Workplace*, pp. 459–479, New York: Marcel Dekker.
BERNSTEIN, I. L., BERNSTEIN, D. I., CHAN-YEUNG, M. & MALO, J.-L. (1993) Definition and classification of asthma, in BERNSTEIN, I. L., CHAN-YEUNG, M., MALO, J.-L. & BERNSTEIN, D. I. (Eds) *Asthma in the Workplace*, pp. 1–4. New York: Marcel Dekker.
BERNSTEIN, J. A. (1992) Occupational asthma. *Postgraduate Medicine*, **92**, 109–118.
BERNSTEIN, J. A. & BERNSTEIN, I. L. (1994) Clinical aspects of respiratory hypersensitivity to chemicals, in DEAN, J. H., LUSTER, M. I., MUNSON, A. E. & KIMBER, I. (Eds) *Immunotoxicology and Immunopharmacology*, pp. 617–642, New York: Raven Press.
BERNSTEIN, J. A., STAUDER, T., BERNSTEIN, D. I. & BERNSTEIN, I. L. (1994) A combined respiratory and cutaneous hypersensitivity syndrome induced by work exposure to quaternary amines. *Journal of Allergy and Clinical Immunology*, **94**, 257–259.
BERNSTEIN, J. A., BERNSTEIN, D. I. & BERNSTEIN, I. L. (1996) Occupational asthma, in BIERMAN, C. W., PEARLMAN, D. S., SHAPIRO, G. G. & BUSSE,

W. W. (Eds) *Allergy, Asthma and Immunology from Infancy to Adulthood*, pp. 529–554. Philadelphia: W. B. Saunders Company.

BIAGINI, R. E., BERNSTEIN, I. L., GALLAGHER, J. S., MOORMAN, W. J., BROOKS, S. & GANN, P. H. (1985) The diversity of reaginic immune responses to platinum and palladium metallic salts. *Journal of Allergy and Clinical Immunology*, **76**, 794–802.

BIAGINI, R. E., KLINCEWICZ, S. L., HENNINGSEN, G. M., MACKENZIE, B. A., GALLAGHER, J. S., BERNSTEIN, D. I. & BERNSTEIN, I. L. (1990) Antibodies to morphine in workers exposed to opiates at a narcotic manufacturing facility and evidence for similar antibodies to heroin abusers. *Life Sciences*, **47**, 897–908.

BIAGINI, R. E., HENNINGSEN, G. M. & KLINCEWICZ, S. L. (1995) Immunologic analyses of peripheral leukocytes from workers at an ethical narcotics manufacturing facility. *Archives of Environmental Health*, **50**, 7–12.

BIGNON, J. S., ARON, Y., YA JU, L., KOPFERSCHMITT, M. C., GARNIER, R., MAPP, C., FABBRI, L. M., PAULI, G., LOCKHART, A. & CHARRON, D. (1994) HLA class II alleles in isocyanate-induced asthma. *American Journal of Respiratory and Critical Care Medicine*, **149**, 71–75.

BLAKE, B. L., MACKAY, J. B., RAINEY, H. B. & WESTON, W. J. (1965) Pulmonary opacities resulting from diisocyanate exposure. *Australian Radiology*, **9**, 45–48.

BROOKS, S. M. (1994) Host susceptibility to indoor air pollution. *Journal of Allergy and Clinical Immunology*, **94**, 344–351.

BROOKS, S. M. & BERNSTEIN, I. L. (1993) Reactive airways dysfunction syndrome or irritant-induced asthma, in BERNSTEIN, I. L., CHAN-YEUNG, M., MALO, J.-L. & BERNSTEIN, D. I. (Eds) *Asthma in the Workplace*, pp. 533–549. New York: Marcel Dekker.

BROOKS, S. M. & VANDERVORT, R. (1977) Polyvinyl chloride film thermal decomposition products as an occupational illness. *Journal of Occupational Medicine*, **18**, 192–196.

BROOKS, S. M., WEISS, M.A. & BERNSTEIN, I. L. (1985) Reactive airways dysfunction syndrome. *Journal of Occupational Medicine*, **27**, 473–476.

BURGE, P. S. (1982a) Occupational asthma due to soft soldering fluxes containing colophony (rosin, pine resin). *European Journal of Respiratory Disease*, **63**, 65–67.

(1982b) Occupational asthma in electronics workers caused by colophony fumes: follow-up of affected workers. *Thorax*, **37**, 348–353.

(1989) Occupational risks of glutaraldehyde. *British Medical Journal*, **299**, 1451.

BURGE, P. S., EDGE, G., HAWKINS, R., WHITE, V. & TAYLOR, A. N. (1981) Occupational asthma in a factory making flux-cored solder containing colophony. *Thorax*, **36**, 828–834.

BURROWS, D. (1984) The dichromate problem. *International Journal of Dermatology*, **23**, 215–220.

BUTCHER, B. T., MAPP, C. E. & FABBRI, L. M. (1993) Polyisocyanates and their prepolymers, in BERNSTEIN, I. L., CHAN-YEUNG, M., MALO, J.-L. & BERNSTEIN, D. I. (Eds) *Asthma in the Workplace*, pp. 415–437. New York: Marcel Dekker.

CARTIER, A. & MALO, J.-L. (1993) Occupational challenge tests, in BERNSTEIN, I. L., CHAN-YEUNG, M., MALO, J.-L. & BERNSTEIN, D. I. (Eds) *Asthma in the Workplace*, pp. 215–248. New York: Marcel Dekker.

CHAN-YEUNG, M. (1982) Immunologic and nonimmunologic mechanisms in asthma due to western red cedar (*Thuja plicata*). *Journal of Allergy and Clinical Immunology*, **70**, 32–37.
CHAN-YEUNG, M. (1993) Western red cedar and other wood dusts, in BERNSTEIN, I. L., CHAN-YEUNG, M., MALO, J.-L. & BERNSTEIN, D. I. (Eds) *Asthma in the Workplace*, pp. 503–531. New York: Marcel Dekker.
CHAN-YEUNG, M. & LAM, S. (1986) State of art: occupational asthma, *American Review of Respiratory Disease*, **133**, 686–703.
CHAN-YEUNG, M., BARTON, G., MACLEAN, L. & GRZYBOWSKI, S. (1973) Occupational asthma and rhinitis due to western red cedar (*Thuja plicata*). *American Review of Respiratory Disease*, **108**, 1094–1102.
CHAN-YEUNG, M., MACLEAN, L. & PAGGIARO, P. L. (1987) Follow-up study of 232 patients with occupational asthma caused by western red cedar (*Thuja plicata*). *Journal of Allergy and Clinical Immunology*, **79**, 792–796.
CHAN-YEUNG, M., LERICHE, J., MACLEAN, L. & LAM, S. (1988) Comparison of cellular and protein changes in bronchial lavage fluid of symptomatic and asymptomatic patients with red cedar asthma on follow-up examination. *Clinical Allergy* **18**, 359–365.
CHAN-YEUNG, M., CHAN, H., SALARI, H. & LAM, S. (1989) Histamine and leukotriene release in bronchial fluid during plicatic acid-induced bronchoconstriction. *Journal of Allergy and Clinical Immunology*, **84**, 762–768.
COOPER, J. A. D., JR., WHITE, D. A. & MATTHAY, R. A. (1986a) Drug-induced pulmonary disease. Part 1: cytotoxic drugs. *American Review of Respiratory Disease*, **133**, 321–340.
(1986b) Drug-induced pulmonary disease. Part 2: noncytotoxic drugs. *American Review of Respiratory Disease*, **133**, 488–506.
DEARMAN, R. J., SPENCE, L. M. & KIMBER, I. (1992) Characterization of murine immune responses to allergenic diisocyanates. *Toxicology and Applied Pharmacology*, **112**, 190–197.
DEMONCHY, G. R., KAUFFMAN, H. F., VENGE, P., KOETER, G. H., JANSEN, H. M., SLUITER, H. J. & DE VRIES, K. (1985) Bronchoalveolar eosinophilia during allergen-induced late asthmatic reactions. *American Review of Respiratory Disease*, **131**, 373–376.
DOCKER, A., WATTIE, J. M., TOPPING, M. D., LUCZYNSKA, C. M., NEWMAN-TAYLOR, A. J., PICKERING, C. A. C. & GOMPERTZ, P. T. (1987) Clinical and immunological investigations of respiratory disease in workers using reactive dyes. *British Journal of Industrial Medicine*, **44**, 534–541.
DURHAM, S. R., GRANECK, J., HAWKINS, R. & NEWMAN-TAYLOR, A. J. (1987) The temporal relationship between increases in airway responsiveness to histamine and late asthmatic responses induced by occupational agents. *Journal of Allergy and Clinical Immunology*, **79**, 398–406.
ENARSON, D. A. & CHAN-YEUNG, M. (1990) Characterization of health effects of wood dust exposures, *American Journal of Industrial Medicine*, **17**, 33–38.
ERASMUS, L. D. (1957) Scleroderma in gold miners on the Witwatersrand with particular reference to pulmonary manifestations. *South African Journal of Clinical Laboratories*, **3**, 209–231.
FABBRI, L. M., MAESTRELLI, P., SAETTA, M. & MAPP, C. E. (1993) Pathophysiology of occupational asthma, in BERNSTEIN, I. L., CHAN-YEUNG, M.,

MALO, J.-L. & BERNSTEIN, D. I. (Eds) *Asthma in the Workplace*, pp. 61–92. New York: Marcel Dekker.

FIEDLER, N., MACCIA, C. & KIPEN, H. (1992) Evaluation of chemically sensitive patients. *Journal of Medicine*, **34**, 529–538.

FINOTTO, S., FABBRI, L. M., RADO, V., MAPP, C. E. & MAESTRELLI, P. (1991) Increase in numbers of CD8 positive lymphocytes and eosinophils in peripheral blood of subjects with late asthmatic reactions induced by toluene diisocyanate. *British Journal of Industrial Medicine*, **48**, 116–121.

GALLAGHER, J. S., TSE, C. S. T., BROOKS, S. M. & BERNSTEIN, I. L. (1981) Diverse profiles of immunoreactivity in toluene diisocyanate (TDI) asthma. *Journal of Occupational Medicine*, **23**, 610–616.

GAUTRIN, D., BOULET, L.-P., BOULET, M., DUGAS, M., BHERER, L., L'ARCHEVEQUE, J., LAVIOLETTE, M., COTE, J. & MALO, J.-L. (1994) Is reactive airways dysfunction syndrome a variant of occupational asthma? *Journal of Allergy and Clinical Immunology*, **93**, 12–22.

GHEYSENS, B., AUWERX, J., VAN DEN EECKKOUT, A. & DEMEDTS, M. (1985) Cobalt induced occupational asthma in diamond polishers. *Chest*, **88**, 740–744.

HERZOG, C. H., VILLIGER, B. & BRAUN, P. (1991) Nickel-specific T cell clones in asthma – preferential use of V-beta 14 in T cell receptor beta chain. *European Respiratory Review*, **4**, A4255.

HUDNELL, H. K., OTTO, D. A., HOUSE, D. E. & MOLHAVE, L. (1992) Exposure of humans to a volatile organic mixture. II. Sensory. *Archives of Environmental Health*, **47**, 31–38.

JAAKKOLA, J. J. K., TUOMAALA, P. & SEPPANEN, O. (1994) Textile wall materials and sick building syndrome. *Archives of Environmental Health*, **49**, 175–181.

KENNEDY, A. L. & BROWN, W. E. (1992) Isocyanates and lung disease: experimental approaches to molecular mechanisms. *Occupational Medicine*, **7**, 301–329.

KESKINEN, H. (1983) Epidemiology of occupational lung diseases: asthma and allergic alveolitis, in KERR, J. W. & GANDERTON, M. A. (Eds) *XIth International Congress of Allergology and Clinical Immunology: Proceedings of invited symposia*, pp. 403–407. London: Macmillan Press Ltd.

KOREN, H. S., GRAHAM, D. E. & DEVLIN, R. B. (1992) Exposure of humans to a volatile organic mixture. III. Inflammatory response. *Archives of Environmental Health*, **47**, 39–44.

KRIEBEL, D., BRAIN, J. D., SPRINCE, N. L. & KAZEMI, H. (1988) The pulmonary toxicity of beryllium. *American Review of Respiratory Disease*, **137**, 464–473.

KUSAKA, Y., NAKONA, Y., SHIRAKAWA, T. & MORIMOTO, K. (1989) Lymphocyte transformation with cobalt in hard metal asthma. *Industrial Health*, **27**, 155–163.

LAX, M. B. & HENNEBERGER, P. K. (1995) Patients with multiple chemical sensitivities in an occupational health clinic: presentation and follow-up. *Archives of Environmental Health*, **50**, 425–431.

LISS, G. M., BERNSTEIN, D. I., GENESOVE, L., ROOS, J.O. & LIM, J. (1993) Assessment of risk factors for IgE mediated sensitization to tetrachlorophthalic anhydride. *Journal of Allergy and Clinical Immunology*, **92**, 237–247.

LUCZYNSKA, C. M., HUTCHCROFT, B. J., HARRISON, M. A., DORNAN, J. D. & TOPPING, M. D. (1990) Occupational asthma and specific IgE to diazonium salt intermediate used in the polymer industry. *Journal of Allergy and Clinical Immunology*, **85** 1076–1082.

Luo, J. D., Nelsen, K. G. & Fischbein, A. (1990) Persistent reactive airway dysfunction syndrome after exposure to toluene diisocyanate. *British Journal of Industrial Medicine*, **47**, 239–241.

McCloud, T. C. (1991) Occupational lung disease. *Radiologic Clinics of North America*, **29**, 931–941.

McKerrow, E. B., McDermott, M., Gilson, J. C. & Schilling, R. S. F. (1958) Respiratory function during the day in cotton workers: a study in byssinosis. *British Journal of Industrial Medicine*, **15**, 75–83.

Malo, J.-L. & Bernstein, I. L. (1993) Other chemical substances causing occupational asthma, in Bernstein, I. L., Chan-Yeung, M., Malo, J.-L. & Bernstein, D. I. (Eds) *Asthma in the Workplace*, pp. 481–502. New York: Marcel Dekker.

Malo, J.-L. & Zeiss, C. R. (1982) Occupational hypersensitivity pneumonitis after exposure to diphenylmethane diisocyanate. *American Review of Respiratory Disease*, **125**, 113–116.

Mapp, C. E., Boschetto, P., Dal Vecchio, L., Maestrelli, P. & Fabbri, L. M. (1988) Occupational asthma due to isocyanates. *European Respiratory Journal*, **1**, 273–279.

Merchant, J. A. & Bernstein, I.L. (1993) Cotton and other textile dusts, in Bernstein, I. L., Chan-Yeung, M., Malo, J.-L. & Bernstein, D. I. (Eds) *Asthma in the Workplace*, pp. 551–576. New York: Marcel Dekker.

Meredith, S. K., Taylor, V. M. & McDonald, J. C. (1991) Occupational respiratory disease in the United Kingdom 1989: a report to the British Thoracic Society and the Society of Occupational Medicine by the SWORD project group. *British Journal of Industrial Medicine*, **48**, 292–298.

Moller, D. R., McKay, R. T., Bernstein, I. L. & Brooks, S. M. (1986a) Persistent airways disease caused by toluene diisocyanate. *American Review of Respiratory Disease*, **134**, 175–176.

Moller, D. R., Brooks, S. M., Bernstein, D. I., Cassedy, K., Enrione, M. & Bernstein, I. L. (1986b) Delayed anaphylactoid reaction in a worker exposed to chromium. *Journal of Allergy and Clinical Immunology*, **77**, 451–456.

Nemery, B. (1990) Metal toxicity and the respiratory tract. *European Respiratory Journal*, **3**, 202–219.

Paggiaro, P. L. & Chan-Yeung, M. (1987) Pattern of specific airway response in asthma due to western red cedar (*Thuja plicata*): relationship with length of exposure and lung function measurements. *Clinical Allergy*, **17**, 333–339.

Patterson, R., Zeiss, C. R. & Pruzansky, J. J. (1982) Immunology and immunopathology of trimellitic anhydride pulmonary reactions. *Journal of Allergy and Clinical Immunology*, **70**, 19–23.

Pauli, G., Bessot, J. C., Kopferschmitt, M. C., Lingot, G., Wendling, R., Ducos, P. & Limasset, J. C. (1980) Meat wrapper's asthma: identification of the causal agent. *Clinical Allergy*, **10**, 263–269.

Pepys, J., Longbotton, J. L., Hargreave, F. E. & Faux, J. (1969) Allergic reactions of the lungs to enzymes of *Bacillus subtilis*. *Lancet*, **i**, 1811–1814.

Pepys, J., Hutchcroft, B. J. & Breslin, A. B. X. (1976) Asthma due to inhaled chemical agents – persulfate salts and henna in hairdressers. *Clinical Allergy*, **6**, 399–404.

PETERSON, W. F. (1946) *Hippocratic Wisdom*, p. 223. Springfield, IL: Charles C. Thomas.

PLAVEC, D., SOMOGYI-ZALUD, E. & GODNIC-CVAR, J. (1993) Nonspecific nasal responsiveness in workers occupationally exposed to respiratory irritants. *American Journal of Industrial Medicine*, **24**, 525–532.

RICHELDI, L., SORRENTINO, R. & SALTINI, C. (1993) HLA-DBB1 glutamate 69: a genetic marker of beryllium disease. *Science*, **262**, 242–244.

ROSSMAN, M. D., KERN, J. A., ELIAS, J. A., CULLEN, M. R., EPSTEIN, P. E., PREUSS, O. P., MARKHAM, T. M. & DANIELLE, R. P. (1988) Proliferative response of bronchoalveolar lymphocytes to beryllium. *Annals of Internal Medicine*, **108**, 687–693.

SAETTA, M., DI STEFANO, A., MAESTRELLI, P., DE MARZO, M., MILANI, G. F., PIVIROTTO, F., MAPP, C. E. & FABBRI, L. M. (1992a) Airway mucosal inflammation in occupational asthma induced by toluene diisocyanate. *American Review of Respiratory Disease*, **145**, 160–168.

SAETTA, M., MAESTRELLI, P., DI STEFANO, A., DE MARZO, N., MILANI, G. F., PIVIROTTO, F., MAPP, C. E. & FABBRI, L. M. (1992b) Effect of cessation of exposure to toluene diisocyanate (TDI) on bronchial mucosa of subjects with TDI-induced asthma. *American Review of Respiratory Disease*, **145**, 169–174.

SALTINI, C., WINESTOCK, K., KIRBY, M., PINKSTON, P. & CRYSTAL, R. G. (1989) Maintenance of alveolitis in patients with chronic beryllium disease by beryllium-specific helper T cells. *New England Journal of Medicine*, **320**, 1103–1109.

SELIGMAN, P. J. & BERNSTEIN, I. L. (1993) Medicolegal and compensation aspects, in BERNSTEIN, I. L., CHAN-YEUNG, M., MALO, J.-L. & BERNSTEIN, D. I. (Eds) *Asthma in the Workplace*, pp. 323–340. New York: Marcel Dekker.

SELTZER, J. M. (1994) Building-related illnesses. *Journal of Allergy and Clinical Immunology*, **94**, 351–361.

SETA, J. A., YOUNG, R. O., BERNSTEIN, I. L. & BERNSTEIN, D. I. (1993) The United States National Exposure Survey (NOES) Data Base, in BERNSTEIN, I. L., CHAN-YEUNG, M., MALO, J.-L. & BERNSTEIN, D. I. (Eds) *Asthma in the Workplace*, pp. 626–663. New York: Marcel Dekker.

SHARMA, O. P. (1991) Hypersensitivity pneumonitis, in BONE, R. C. (Ed) *Disease-a-Month*, pp. 411–471. St. Louis, MO: Mosby-Year Book, Inc.

SHEPPARD, D., THOMPSON, J. E., SCYPINSKI, L., DRUSSER, D., NADEL, J. A. & BORSON, D. B. (1988) Toluene diisocyanate increases airway responsiveness to substance P and decreases airway neutral endopeptidase. *Journal of Clinical Investigation*, **81**, 1111–1115.

SHIRAKAWA, T., KUSAKA, Y., FUKIMURA, N., GOTO, S. & MORIMOTO, K. (1989) Occupational asthma from cobalt sensitivity in workers exposed to hard metal dust. *Chest*, **95**, 29–37.

SIMON, G. E., DANIELL, W., STOCKBRIDGE, H., CLAYPOOLE, K. & ROSENSTOCK, L. (1993) Immunologic, psychological and neuropsychological factors in multiple chemical sensitivity. *Annals of Internal Medicine*, **19**, 97–103.

TARLO, S. M. & BRODER, I. (1989) Irritant-induced occupational asthma. *Chest*, **96**, 297–300.

THOMPSON, J. E., SCYPINSKI, L., GORDON, T. & SHEPPARD, D. (1987) Tachykinins mediate the acute increase in airway responsiveness caused by toluene diisocyanate in guinea pigs. *American Review of Respiratory Disease*, **136**, 43–49.

THORPE, J., STEINBERG, D. R., BERNSTEIN, I. L. & MURLAS, C. (1987) Bronchial reactivity increases soon after the immediate response in dual-responding asthma subjects. *Chest*, **91**, 21–25.
VANDERVORT, R. & BROOKS, S. M. (1977) Polyvinyl chloride film thermal decomposition products as an occupational illness. I. Environmental and experimental toxicology. *Journal of Occupational Medicine*, **19**, 188–191.
VENABLES, K. M., TOPPING, M. D., HOWE, W., LUCZYNSKI, C. M., HAWKINS, R. & NEWMAN-TAYLOR, A. J. (1985) Interaction of smoking and atopy in producing specific IgE antibody against a hapten protein conjugate. *British Medical Journal*, **290**, 201–204.
WEGNER, C. D., JACKSON, A. C., BERRY, J. D. & GILLESPIE, J. R. (1984) Intercellular adhesion molecule-1 (ICAM-1) in the pathogenesis of asthma. *Science*, **247**, 456–459.
ZEISS, C. R. & PATTERSON, R. (1993) Acid anhydrides, in BERNSTEIN, I. L., CHAN-YEUNG, M., MALO, J.-L. & BERNSTEIN, D. I. (Eds) *Asthma in the Workplace*, pp. 439–457. New York: Marcel Dekker.
ZEISS, C. R., KKANELLAKES, T. M. & BELLONE, J. D. (1980) Immunoglobulin E-mediated asthma and hypersensitivity pneumonitis and precipitating antihapten antibodies due to diphenylmethane diisocyanate. *Journal of Allergy and Clinical Immunology*, **65**, 345–352.
ZOCCA, E., FABBRI, L. M., BOSHETTO, P., MASIERO, M., MILANI, G. F., PIVIROTTO, P. & MAPP, C. E. (1990) Leukotriene B4 and late asthmatic reactions induced by toluene diisocyanate. *Journal of Applied Physiology*, **68**, 1576–1589.

4

Occupational Respiratory Allergy: Risk Assessment and Risk Management

KATHERINE M. VENABLES and ANTHONY J. NEWMAN TAYLOR

Department of Occupational and Environmental Medicine, Imperial College School of Medicine at the National Heart and Lung Institute, London

4.1 Definitions of Terms

The focus of this chapter is occupational asthma mediated by immunological hypersensitivity, termed 'occupational asthma with a latency period' by Bernstein et al. (1993). This may involve formation of IgE antibody, but not necessarily. Rhinitis, conjunctivitis and urticaria are included because they co-exist commonly with asthma and share common mechanisms. Occupational asthma caused by an irritant exposure sufficient to damage airway mucosa is excluded. There may be immunological epiphenomena, such as formation of specific antibody, in irritant-induced asthma but, when present, these do not imply an immunological causal pathway.

Risk assessment means an evaluation of the likelihood that occupational asthma will occur. Risk management uses risk assessments to inform decision-making about prevention of occupational asthma. These processes are carried out within individual workplaces, across industrial sectors (as happened after the worldwide epidemics of asthma in enzyme detergent manufacturing workers (Juniper et al., 1977)), by governmental regulatory agencies such as the Health and Safety Executive in the UK, and international advisory bodies (for example Vos et al., 1996 on behalf of the World Health Organization).

4.2 Aspects of the Biology of Occupational Asthma Relevant to Risk Assessment and Management

4.2.1 Exposure

There are several hundred well-documented occupational causes of asthma

(Chan-Yeung and Malo, 1993). They fall into two groups: chemicals, the subject of this book, such as isocyanates or acid anhydrides, and complex biological materials, such as rat urine or flour.

There is little information on exposure–response relationships for the induction of asthma. There are several explanations for this. Estimates of workers' chemical exposure are needed, which usually require a sampling survey of representative current exposures and assessment of past exposure. Making use of measurements made to document compliance with regulations is rarely sufficient because such measurements generally focus only on those jobs and tasks likely to cause violations of occupational standards.

Information on representative exposures must be linked to information about cases of occupational asthma, which should be as complete as possible. However, this is a new area for research and most exposure–response analyses so far have come from research which is likely to be seriously deficient: the number of cases is underestimated or they are unrepresentative of cases in general or both. This is because most research has used prevalence measures (number of cases present/number of workers at risk) within dynamic population cross-sections (those working in a workplace at one point in time). This approach yields data rapidly but is at risk of survivor bias. Workers with respiratory symptoms related to their work are much more likely to leave their job than other workers and it is likely that this selective turnover operates most strongly in the most highly exposed jobs.

Many cross-sectional surveys report no or weak exposure–response relationships or even an inverse relationship for which survivor bias is the most probable explanation. For example, in a survey of a workforce exposed to tetrachlorophthalic anhydride (TCPA) carried out by the authors, all the cases affected in an epidemic of occupational asthma had left the factory before it was possible to carry out a survey (Venables *et al.*, 1985a). In this survey, conclusions could be drawn by studying specific IgE against TCPA. However, reliable tests for specific IgE antibody are not available for all chemical respiratory allergens.

It may require considerable resources to assemble a complete case register covering a defined period of time. Usually this entails a systematic survey of workers who have left because few occupational health services would aim at collecting such data routinely. Tracing leavers is more difficult and time-consuming than studying workers still in employment.

Once data are available, several types of analysis are possible. The most straightforward is comparison of disease incidence rate in different exposure groups (number of new cases/person-time). The case-referent approach is conceptually similar but will yield risk multipliers rather than direct measurements of incidence rate. Comparisons of prevalence may be problematic because of the survivorship issues discussed above.

Exposure–response analyses published so far have been qualitative, with exposure categorized as 'high', 'medium' and 'low' or in some similar form. Such analyses have shown that the risk of asthma (or of some outcome measure closely associated with asthma) increases with exposure in, for example, enzyme detergent workers (Juniper et al., 1977) and platinum refinery workers (Calverley et al., 1995). The next step will be quantitative exposure–response modelling to explore the shape of the relationships, if threshold or ceiling exposures are present, and the magnitude of the increase in response with each increment in exposure. This is a contentious area involving many assumptions with analogies to the discussion surrounding quantitative estimates of cancer risk associated with radiation.

Animal models may provide useful insights into exposure–response relationships. For example, animal models using subtilisin and isocyanates have shown an increase in percentage of guinea pigs sensitized by inhalation with increasing concentration of sensitizing agent (Karol et al., 1985). The animal model of asthma caused by subtilisin suggests that short-term exposure to a high level of enzyme was a more effective sensitizer than exposure to an equivalent total dose at low levels over a long time period (Thorne et al., 1986).

4.2.2 Smoking

Personal smoking increases the risk in some chemical exposures. This was first observed when IgE antibody against TCPA was noted to be more common in the serum of smokers than in non-smokers (Venables et al., 1985a). In a historical cohort study, smoking platinum refinery workers were found to be at five times greater risk of a positive skin prick test to complex platinum salts than were non-smokers (Venables et al., 1989a). Calverley et al. (1995) have extended this observation by showing that smoking increases the risk of sensitization to platinum salts in those workers with high exposure to complex platinum salts (Figure 4.1). The increased risk in smokers is seen in occupational sensitization by some high-molecular-weight allergens as well (Lancet, 1985) and a rodent model of airway sensitization has confirmed this (Zetterstrom et al., 1985).

Suggested explanations include increased airway mucosal transport in smokers and modification of immune regulatory pathways. It may be that other irritant gases and vapours can have a similar effect. There have been several reports of enhancement of experimental sensitization in animals by irritants (Matsumura, 1970; Osebold et al., 1980). These include ozone enhancement of sensitization by complex platinum salts in a primate model (Biagini et al., 1986).

It is puzzling that this risk from smoking is not seen in all chemical exposures that cause occupational asthma. For example, a large series of 232 patients with asthma caused by western red cedar wood (in which the causal agent is believed to be plicatic acid) contained only 5 per cent of current smokers (Chan-Yeung et al., 1987). This is similar to isocyanate-induced

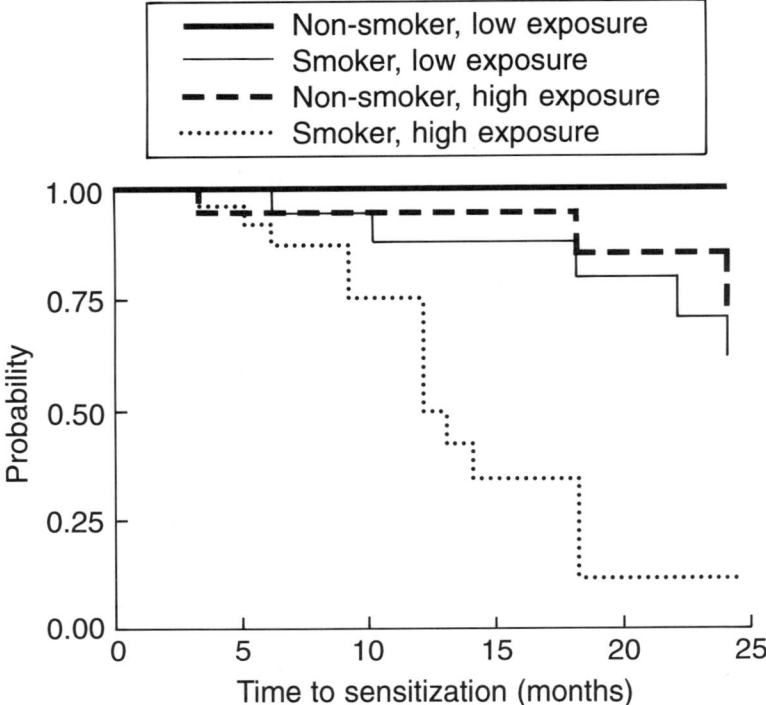

Figure 4.1. Probability of platinum refinery workers remaining skin prick test negative to complex platinum salts, by smoking and exposure categories (reproduced from Calverley et al., 1995, with permission).

asthma, raising the possibility that enhancement by smoking may be a marker of an IgE-mediated mechanism.

4.2.3 Genetic Background

Allergic hypersensitivity to common environmental allergens tends to cluster in families so it is reasonable to assume that genetic background might predispose to occupational asthma. So far, this has mainly been studied in relation to 'atopy'. This is a nebulous concept named by Coca and Cooke (1923) from the Greek for strange or unusual. In research and clinical practice, it is usually documented using skin prick tests to common allergens, which extends the confusion because it mixes susceptibility with response. Despite these imprecisions, in some exposures atopic individuals are at increased risk. This has been noted for TCPA (Venables et al., 1985a) and complex platinum salts (Venables et al., 1989a), but the effect is weaker than that of smoking. Atopy is a strong risk factor for enzyme-induced asthma (Juniper et al., 1977). It is not a risk factor for isocyanate-induced asthma.

It is now recognized that strains of animals can be bred with either a high or low IgE response to sensitization and the high responders develop tolerance to antigen less readily (Holt and Sedgewick, 1987). However, atopy is not the human equivalent of a high IgE responder animal strain. It is likely that the concept incorporates many genes and that genetic susceptibility to occupational asthma may be complex. For example, Young et al. (1995) suggested that HLA-DR3 increases the risk of sensitization by trimellitic anhydride (TMA) and possibly TCPA but not the closely related phthalic anhydride (PA). Such findings require confirmation before they can be generalized.

4.3 General Principles of Prevention

Prevention is usually conceptualized under three headings: primary, secondary and tertiary. Primary prevention is control of the cause of disease, secondary prevention is early detection of disease coupled with action to minimize its severity, and tertiary prevention is application of appropriate health care to minimize complications in established disease. General examples are in Table 4.1.

4.3.1 Primary Prevention

The most direct method of preventing occupational asthma is to eliminate or reduce the exposures that cause it, or to substantially reduce the size of the exposed population. Corn (1983) has provided a useful list of methods to control exposure to allergens following general principles of occupational hygiene practice. A version of this list is presented in Table 4.2. Usually, a prevention programme will use several of these methods. After the epidemics of occupational asthma due to enzymes in detergents in the 1960s and 1970s, enzyme was handled as a slurry rather than a dust; it was encapsulated; some of the plant was enclosed; exhaust ventilation was increased; protective clothing was modified and education and administrative systems were set up (Juniper et al., 1977). A similar package of multiple measures, with particular reliance on enclosure of processes in glove-boxes, was used successfully in platinum refining (Hughes, 1980). Corn (1983) also included health screening in his list but this does not reduce exposure to the causal agent. In both the enzyme detergent and platinum refinery examples, the preventive measures included screening of prospective employees to identify those judged to be especially susceptible and exclude them from exposure. However, this practice is based on beliefs about susceptibility rather than on good evidence from prospective studies and cannot be supported.

Table 4.1 Examples of prevention strategies in the general population and for occupational asthma caused by chemicals

Type of prevention strategy	General public health	Occupational asthma caused by chemicals
Primary	Smoking reduction programmes to control lung cancer, ischaemic heart disease, chronic bronchitis and other diseases	Environmental control of chemicals that cause asthma
	Road accident control programmes involving traffic management, street lighting, driver education, random breath tests and other measures	
Secondary	Cervical screening to control cancer by early detection of pre-malignancy, removal of dysplastic tissue and monitoring for recurrence	Screening programmes to detect allergy to a chemical, or asthma at an early stage, coupled with action to reduce or eliminate exposure so as to prevent progression
	Routine measurement of blood pressure to detect hypertension and treat affected people with the aim of preventing stroke and other disease	
Tertiary	Active rehabilitation after stroke to improve speech and mobility and prevent pressure sores and other complications	Treatment of asthma by reducing or eliminating exposure, with drug treatment if necessary, to prevent complications such as death

In exposures where smoking is a risk factor for occupational asthma, education and other measures to control smoking should reduce disease risk by reducing the number of smokers. It may also suggest that exposure to chemical irritants should be controlled in the workplace. Knowledge of the genetics of chemical-induced occupational asthma is so limited at present that no useful measures can be taken to identify susceptible workers and remove them from exposure.

Table 4.2 Methods for controlling exposure to chemical sensitizers in the workplace

Elimination
Substitution
Isolation
Enclosure
Ventilation
Process change
Product change
General housekeeping
Dust suppression
Maintenance
Sanitation
Work practices
Personal protective devices
Waste disposal practices
Administrative controls
Education
Labelling and warning
Environmental monitoring
Management programmes

(after Corn, 1983)

4.3.2 Secondary Prevention

The long-term outcome in occupational asthma after removal from exposure varies considerably from patient to patient, but frequently mild to moderately severe chronic asthma persists after avoidance of exposure to the initiating cause. Follow-up studies have been undertaken in cases of asthma caused by several chemicals, for example western red cedar (Chan-Yeung et al., 1982), solder fumes (Burge, 1982), isocyanates (Paggiaro et al., 1984) and TCPA (Venables et al., 1987); all show persistent respiratory symptoms and non-specific bronchial hyperresponsiveness. This functional impairment is an important factor in the depression, unemployment and social disablement experienced by patients (Venables et al., 1989b). There is some evidence that early detection of occupational asthma coupled with prompt removal from exposure improves this prognosis (Chan-Yeung et al., 1982). This is the rationale for screening programmes during employment.

Screening for occupational asthma is usually carried out using a symptoms questionnaire and individual occupational health services may supplement this by lung function testing and, where appropriate, skin prick or other immunological tests. Such screening programmes can use considerable resources without clear objectives (Newill et al., 1986). It is important for them to be coupled with a clear policy on the management of abnormal tests discovered during screening so that cases of occupational asthma are diagnosed without delay. Few programmes are linked to some form of audit of the efficiency and effectiveness of the screening process.

4.3.3 Tertiary Prevention

Occupational asthma is potentially fatal. In two recent reports of deaths, the patients had remained exposed to the causal agents, TDI and flour, for economic reasons (Fabbri et al., 1988; Ehrlich, 1994). Care of a worker with a diagnosis of occupational asthma must therefore focus on eliminating exposure. Unfortunately, this often entails loss of employment. Patients who decide to remain in a job with exposure against medical advice can be supported by respiratory protection and medication but this must be accompanied by monitoring of symptoms, lung function and, where applicable, immunological responsiveness.

4.4 Risk Assessment

The first element in risk assessment is knowledge of the chemical exposures of a workforce to see if a known sensitizer is present. A chemical becomes accepted as a cause of occupational asthma if there are at least two convincing clinical case reports from independent centres. Specialist clinicians are likely to be aware of the published literature at an early stage and may be a useful source of informal information. There is currently much interest in publishing lists of sensitizers. One is available in a recent textbook (Chan-Yeung and Malo, 1993). The Health and Safety Executive in the UK, the National Institute for Occupational Safety and Health in the USA and other agencies may publish lists with updates. Other sources of information on causes are reporting schemes such as SWORD (Surveillance of Work-related and Occupational Respiratory Disease) in the UK (Ross et al., 1995) and SENSOR (Sentinel Event Notification System for Occupational Risks) in some states in the USA (Matte et al., 1990).

Even if there are no known sensitizers, occupational health staff should be alert to the possibility of occupational asthma and informed about changes to raw materials and manufacturing processes. Workers with new asthma, or significant exacerbation of pre-existing asthma, should be assessed carefully. Often, one or two cases of work-related asthma are only the tip of the iceberg. For example, investigations at a steel coating plant with two cases of work-related asthma suggested there were at least 20 cases of asthma caused by TDI, which had been introduced into the process by a supplier (Venables et al., 1985b). Clusters such as this are worth investigating in detail.

If a known sensitizer is present, the available literature should be consulted with the following questions in mind. Are there published occupational exposure standards? Have governmental agencies or industry-wide working groups published guidelines on control of exposures or working practices? Is there published research on exposure–response relationships? Is there literature on other risk factors? If there is no or scanty information, it would be worth setting up a study with the aim of publication so that others can make use of the findings.

With a known sensitizer, it will be useful to know the current environmental levels of the chemical and the asthma frequency rate in the workforce. If there is no regular environmental monitoring programme, a one-off survey would be helpful. A health surveillance programme to monitor respiratory symptoms will be needed and this may be supplemented by lung function tests and, where applicable, immunological tests.

4.5 Risk Management

This depends on the scale of the problem with occupational asthma. Major industry-wide changes have been made to control asthma due to enzyme detergents and complex platinum salts, for example. Individual factories have changed raw materials, processes and ventilation in response to outbreaks of asthma. In workplaces with no or occasional cases the priority is adherence to the appropriate guidelines on work practices, which may be from a regulatory agency, an industry-wide group, or local committee. The asthma frequency should be monitored by a consistent methodology to evaluate the effectiveness of these controls.

References

BERNSTEIN, I. L., CHAN-YEUNG, M., MALO, J.-L. & BERNSTEIN, D. I. (1993) Definition and classification of asthma, in BERNSTEIN, I. L., CHAN-YEUNG, M., MALO, J.-L. & BERNSTEIN, D. I. (Eds) *Asthma in the Workplace*, pp. 1–4. New York: Marcel Dekker.

BIAGINI, R. E., MOORMAN, W. J., LEWIS, T. R. & BERNSTEIN, I. L. (1986) Ozone enhancement of asthma in a primate model. *American Review of Respiratory Disease*, **134**, 719–725.

BURGE, P. S. (1982) Occupational asthma in electronics workers: follow-up of affected workers. *Thorax*, **37**, 348–353.

CALVERLEY, A. E., REES, D., DOWDESWELL, R. J., LINNETT, P. J. & KIELKOWSKI, D. (1995) Platinum salt sensitivity in refinery workers: incidence and effects of smoking and exposure. *Occupational and Environmental Medicine*, **52**, 661–666.

CHAN-YEUNG, M. & MALO, J.-L. (1993) Table of the major inducers of occupational asthma, in BERNSTEIN, I. L., CHAN-YEUNG, M., MALO, J.-L. & BERNSTEIN, D. I. (Eds) *Asthma in the Workplace*, pp. 595–624. New York: Marcel Dekker.

CHAN-YEUNG, M., LAM, S. & KOENER, S. (1982) Clinical features and natural history of occupational asthma due to western red cedar (*Thuja plicata*). *American Journal of Medicine*, **72**, 411–415.

CHAN-YEUNG, M., MACLEAN, L. & PAGGIARO, P. L. (1987) Follow up study of 232 patients with occupational asthma caused by western red cedar (*Thuja plicata*). *Journal of Allergy and Clinical Immunology*, **79**, 792–796.

COCA, A. F. & COOKE, R. A. (1923) On the classification of the phenomena of hypersensitiveness. *Journal of Immunology*, **8**, 163–182.

CORN, M. (1983) Assessment and control of environmental exposure. *Journal of Allergy and Clinical Immunology*, **72**, 231–241.

EHRLICH, R. I. (1994) Fatal asthma in a baker: a case report. *American Journal of Industrial Medicine*, **26**, 799–802.

FABBRI, L. M., DANIELI, D., CRESCIOLI, S., BEVILACQUA, P., MELI, S., SAETTA, M. & MAPP, C. E. (1988) Fatal asthma in a toluene diisocyanate sensitized subject. *American Review of Respiratory Disease*, **137**, 1494–1498.

HOLT, P. G. & SEDGEWICK, J. D. (1987) Suppression of IgE responses following inhalation of antigen: a natural homeostatic mechanism which limits sensitization to aeroallergens. *Immunology Today*, **8**, 14–15.

HUGHES, E. G. (1980) Medical surveillance of platinum refinery workers. *Journal of the Society of Occupational Medicine*, **30**, 27–30.

JUNIPER, C. P., HOW, M. J., GOODWIN, B. F. J. & KINSHOTT, A. K. (1977) *Bacillus subtilis* enzymes: a 7 year clinical, epidemiological and immunological study of an industrial allergen. *Journal of the Society of Occupational Medicine*, **27**, 3–12.

KAROL, M. H., STADLER, J. & MAGRENI, C. (1985) Immunotoxicologic evaluation of the respiratory system: animal models for immediate- and delayed-onset pulmonary hypersensitivity. *Fundamental and Applied Toxicology*, **5**, 459–472.

LANCET (EDITORIAL) (1985) Smoking, occupation and allergic lung disease. *Lancet*, **i**, 965.

MATSUMURA, Y. (1970) The effects of ozone, nitrogen dioxide, and sulfur dioxide on the experimentally induced allergic respiratory disorder in guinea pigs I: the effect on sensitization with albumin through the airway. *American Review of Respiratory Disease*, **102**, 430–437.

MATTE, T. D., HOFMAN, R. E., ROSEMAN, K. D. & STANDBURY, M. (1990) Surveillance of occupational asthma under the SENSOR model. *Chest*, **98** (Suppl.), 173–178.

NEWILL, C. A., EVANS, R. & KHOURY, M. J. (1986) Pre-employment screening for allergy to laboratory animals: epidemiologic evaluation of its potential usefulness. *Journal of Occupational Medicine*, **28**, 1158–1164.

OSEBOLD, J. W., GERSHWIN, L. J. & ZEE, Y. C. (1980) Studies on the enhancement of allergic lung sensitization by inhalation of ozone and sulfuric acid aerosol. *Journal of Environmental Pathology and Toxicology*, **3**, 221–234.

PAGGIARO, P. L., LOI, A. M., ROSSI, O., FERRANTE, B., PARDI, F., ROSELLI, M. G. & BASCHIERI, L. (1984) Follow-up study of patients with respiratory disease due to toluene di-isocyanate (TDI). *Clinical Allergy*, **14**, 463–469.

ROSS, D. J., SALLIE, B. A. & MCDONALD, J. C. (1995) SWORD '94: surveillance of work-related and occupational respiratory disease in the UK. *Occupational Medicine*, **45**, 175–178.

THORNE, P. S., HILLEBRAND, J., MAGRENI, C., RILEY, E. J. & KAROL, M. H. (1986) Experimental sensitisation to subtilisin I. Production of immediate- and late-onset pulmonary reactions. *Toxicology and Applied Pharmacology*, **86**, 112–123.

VENABLES, K. M., TOPPING, M. D., HOWE, W., LUCZYNSKA, C. M., HAWKINS, R. & NEWMAN TAYLOR, A. J. (1985a) Interaction of smoking and atopy in producing specific IgE antibody against a hapten protein conjugate. *British Medical Journal*, **290**, 201–204.

VENABLES, K. M., DALLY, M. B., BURGE, P. S., PICKERING, C. A. C. & NEWMAN TAYLOR, A. J. (1985b) Occupational asthma in a steel coating plant. *British Journal of Industrial Medicine*, **42**, 517–524.

VENABLES, K. M., TOPPING, M. D., NUNN, A. J., HOWE, W. & NEWMAN TAYLOR, A. J. (1987) Immunologic and functional consequences of chemical (tetrachlorophthalic anhydride)-induced asthma after four years of avoidance of exposure. *Journal of Allergy and Clinical Immunology*, **80**, 212–218.

VENABLES, K. M., DALLY, M. B., NUNN, A. J., STEVENS, J. F., STEPHENS, R., FARRER, N., HUNTER, J. V., STEWART, M., HUGHES, E. G. & NEWMAN TAYLOR, A. J. (1989a) Smoking and occupational allergy in workers in a platinum refinery. *British Medical Journal*, **299**, 939–942.

VENABLES, K. M., DAVISON, A. G. & NEWMAN TAYLOR, A. J. (1989b) Consequences of occupational asthma. *Respiratory Medicine*, **83**, 437–440.

VOS, J. G., YOUNES, M. & SMITH, E. (Eds) (1996) *Allergic Hypersensitivities Induced by Chemicals: Recommendations for Prevention*. WHO Europe, Boca Raton: CRC press.

YOUNG, R. P., BARKER, R. D., PILE, K. D., COOKSON, W. O. C. M. & NEWMAN TAYLOR, A. J. (1995) The association of HLA-DR3 with specific IgE to inhaled acid anhydrides. *American Journal of Respiratory and Critical Care Medicine*, **151**, 219–221.

ZETTERSTROM, O., NORDVALL, S. L., BJORKSTEN, B., AHLSTEDT, S. & STELANDER, M. (1985) Increased IgE antibody responses in rats exposed to tobacco smoke. *Journal of Allergy and Clinical Immunology*, **75**, 594–598.

5

Immunobiology of Chemical Respiratory Sensitization

IAN KIMBER and REBECCA J. DEARMAN

Zeneca Central Toxicology Laboratory, Alderley Park, Macclesfield

5.1 Introduction

It is important to preface consideration of the immunobiological mechanisms that result in the induction and elicitation of chemical respiratory allergy with a definition of allergy itself. Within the context of toxicology, allergy may be defined operationally as the adverse health effects that result from the stimulation by chemicals, drugs or proteins of specific immune responses. This serves to distinguish chemical allergens, those chemicals that cause adverse health effects secondary to the stimulation of an immune (allergic) response, from those agents that may provoke similar symptoms, but via irritant, pharmacological or other non-immune mechanisms. The term allergy is a broad one that embraces a wide variety of immunological responses that in turn provoke different forms of adverse reactions. This is true of occupational respiratory allergic diseases which include extrinsic allergic alveolitis, hypersensitivity pneumonitis, berylliosis and others. However, for the purposes of this chapter attention will focus upon asthma and related symptoms induced by exposure to chemicals. In practice the characteristics of respiratory allergic hypersensitivity are those of specific pulmonary reactions that occur usually in only a proportion of the exposed population. Respiratory symptoms can be provoked in sensitized individuals by atmospheric concentrations of the inducing chemical allergen that previously were tolerated and that fail in non-sensitized individuals to cause similar symptoms (Newman Taylor, 1988).

There is a large number of chemicals that is able to cause allergy in susceptible individuals. Many of these are associated with skin sensitization or allergic contact dermatitis. Contact allergens are chemicals that can gain access to the viable epidermis and that are protein-reactive (or are converted metabolically to protein-reactive species) such that they can form immunogenic complexes and stimulate a specific immune response that results in sensitization.

Chemicals that are known to cause sensitization of the respiratory tract and occupational asthma are also protein-reactive, but far fewer in number. An intriguing question is why some chemicals induce contact sensitization, whereas others preferentially cause respiratory allergy. The possible mechanistic basis for such selectivity is explored in detail later in this chapter. Clearly, however, the nature of immune responses provoked by the inducing chemical is a critical determinant of the form allergic sensitization will take. Although our understanding of the detailed immunobiological events that result in the effective induction of contact sensitization is far from complete, it is nevertheless apparent that contact hypersensitivity reactions are dependent upon cellular immune responses and the action of effector T-lymphocytes. There is far less certainty about the nature of immune responses necessary for chemical respiratory sensitization.

In the case of protein allergens there is no doubt that immediate-type respiratory hypersensitivity reactions are effected by IgE antibody-associated mechanisms. Indeed, the demonstration of specific IgE antibody is an important component of the diagnosis of hay fever and extrinsic allergic asthma. It is believed that exposure of the susceptible individual to the inducing protein allergen results in the stimulation of an IgE antibody response. Such antibodies distribute systemically and associate with mast cells, including mast cells in the respiratory tract. Following subsequent inhalation exposure to the same protein allergen a respiratory hypersensitivity reaction will be elicited. Specific IgE antibodies borne by tissue mast cells are cross-linked by the protein allergen. This results in mast cell degranulation and the release of both preformed and newly synthesized inflammatory mediators that act together to cause vasodilation, bronchoconstriction and the elicitation of respiratory hypersensitivity (Garssen et al., 1996).

It cannot be assumed, however, that respiratory allergy and occupational asthma caused by chemical sensitizers are necessarily mediated by identical mechanisms. This uncertainty arises from the fact that in many instances, particularly in the case of isocyanate-induced occupational asthma, it has proven difficult to demonstrate the presence of specific IgE antibody in symptomatic patients (Chan-Yeung, 1995). Although associations between serum IgE antibody and allergic sensitization of the respiratory tract to chemicals have been reported (Moller et al., 1985; Salvaggio et al., 1988; Drexler et al., 1993; Karol et al., 1994; Nielsen et al., 1994), there is doubt whether there exists a universal obligatory role for IgE in this process (Chan-Yeung, 1990, 1995). It may be that the failure in some cases to identify specific IgE antibodies in the serum of symptomatic patients is attributable to the use of insufficiently sensitive or inappropriate detection methods. Alternatively, mast cell-binding (homocytotropic) antibody of an isotype other than IgE (in humans IgG4) may be the primary or exclusive mediator of respiratory hypersensitivity. There is some evidence to support a role for IgG4 antibody in the pathogenesis of chemical respiratory allergy (Topping et al., 1989; Park and Hong, 1991). It can not be excluded either that cell-mediated immune responses play an important part in the development of respiratory sensitization; certainly, as will be discussed later, T-lymphocytes participate actively in the chronic bronchial inflammatory processes associated with allergic asthma.

Presently it must be concluded that while for some chemical allergens specific IgE antibody plays an important pathogenetic role, particularly in the context of acute-onset responses, in other instances sensitization may develop independently of IgE (Fabbri et al., 1994; Kimber, 1995, 1996a).

Notwithstanding these considerations, an appreciation of IgE antibody and related responses and the immune processes that serve to induce and regulate them is central to an examination of chemical respiratory sensitization.

5.2 Cytokines and IgE Antibody Responses

The stimulation of IgE responses and the continued production of IgE antibody are regulated tightly by the immune system and controlled by cytokines. Of critical importance in the induction and maintenance of IgE antibody is interleukin 4 (IL-4) which causes a switch to, and the increased expression of, Cε germline transcripts (Finkelman et al., 1986, 1988b; Azuma et al., 1987). Mice that lack the gene for IL-4 also lack IgE and fail to produce IgE antibody to stimuli that normally provoke strong responses of this isotype (Kuhn et al., 1991). In contrast, mice that carry an IL-4 transgene, and that constitutively express high levels of this cytokine, display increased serum concentrations of IgE and exhibit more aggressive IgE antibody responses than do wild-type controls (Tepper et al., 1990; Burstein et al., 1991). Reciprocal effects are mediated by another cytokine, interferon-γ (IFN-γ), that antagonizes IgE antibody production (Finkelman et al., 1988a). Such regulation of IgE responses is not restricted to the mouse. In humans also IL-4 and IFN-γ exert reciprocal influences on IgE production in vitro (Del Prete et al., 1988; Pene et al., 1988; Romagnani et al., 1989; Chretien et al., 1990). In the context of humoral immune responses to chemical respiratory allergens it may prove important that IgG4 is subject to an identical, or at least very similar, reciprocal regulation by IL-4, IFN-γ and other cytokines (Ishizaka et al., 1990; Armitage et al., 1993; Aversa et al., 1993; Punnonen et al., 1993a, 1993b). Certainly there is evidence in humans for the coregulation of IgE and IgG4 antibody responses in certain pathological conditions (Hussain and Ottesen, 1986; Hammarstrom and Smith, 1987). Another cytokine also appears to be important in the coinduction of human IgE and IgG4 responses. Interleukin 13 (IL-13), a molecule that is related to, and shares some homology with IL-4, has been shown recently to effect IgE and IgG4 class switching in human B-lymphocytes by IL-4-independent mechanisms (Cocks et al., 1993; Minty et al., 1993; Punnonen et al., 1993a). Like IL-4, IL-13 also induces the expression by B-lymphocytes of CD23, the low affinity membrane receptor for IgE (Punnonen et al., 1993a), that may itself promote IgE production (Bonnefoy et al., 1993).

In the context of cytokine regulation of IgE responses it is worth noting that other molecules may play direct or indirect roles that are possibly of lesser importance. It has been shown, for instance, that interleukin 8 inhibits selectively the synthesis of IgE by IL-4 primed human B lymphocytes through a mechanism that is independent of the action of IFN-γ (Kimata et al., 1992). In addition, recent evidence suggests that interleukin 7 (IL-7), a cytokine produced by stromal cells in the thymus and bone marrow, may indirectly influence IgE responses secondary to differential effects on the production of IL-4 and IFN-γ. It was observed that IL-7 caused a preferential upregulation of IFN-γ expression by activated human T-lymphocytes (Borger et al., 1996).

The stimulation of IgE and IgG4 antibody production during immune responses is dependent therefore upon the relative availability of regulatory cytokines; those of greatest significance being IL-4, IL-13 and IFN-γ. As the immune response develops the balance between these cytokines is a function largely of their differential production by discrete functional subpopulations of T-lymphocytes, both CD4$^+$ and CD8$^+$ cells.

5.3 T-Lymphocyte Subpopulations and the Allergic Response

That there exists functional heterogeneity among CD4$^+$ T helper (Th) cells has been recognized for over ten years. In 1986 Mosmann and his colleagues described two populations of cloned murine Th cells, designated Th1 and Th2, that could be distinguished on the basis of the patterns of cytokines secreted (Mosmann et al., 1986). The significance of this dichotomy for the development of allergic responses is that IL-4 is a product exclusively of Th2 cells, while only Th1 cells produce IFN-γ. Both Th1 and Th2 cells produce granulocyte/macrophage colony-stimulating factor (GM-CSF) and interleukin 3 (IL-3). However, interleukin 2 (IL-2), tumour necrosis factor-β (TNF-β) and IFN-γ are products only of Th1 cells and interleukins 5, 6 and 10 (IL-5, IL-6 and IL-10), together with IL-4, are elaborated only by Th2 cells (Mosmann and Coffman, 1989; Mosmann et al., 1991). The evolutionary and operational significance of these functional subpopulations is undoubtedly in directing the quality of immune responses such as to provide optimal host resistance against infectious microorganisms (Mosmann and Sad, 1996). The functions of Th cells reflect their cytokine products. Th1 cells effect cell-mediated immune and inflammatory reactions, whereas Th2 cells have as their major functional responsibility the promotion of antibody responses, and in particular IgE and IgG4 antibody responses. As will be described later, the products of Th2 cells also promote other aspects and other components of allergic responses.

The heterogeneity among Th cells is considerably more complex than a simple division into two discrete classes. Th1 and Th2 cells must be regarded as being the most differentiated forms of CD4$^+$ cells that develop from a common precursor or precursors as the immune response evolves. The immediate progenitor population is Th cells (designated Th0) that are characterized by the ability to produce both Th1- and Th2-type cytokines. Th0 cells may themselves develop from a more primitive precursor, Thp, that is believed to secrete only IL-2 (Bendelac and Schwartz, 1991; Mosmann et al., 1991; Kamogawa et al., 1993). Even among mature Th1 and Th2 cell populations there may be additional levels of heterogeneity. Individual cells within populations may show considerable variations in the amount of particular cytokines produced and certainly not all, and perhaps few, Th2-type cells display co-ordinate expression of, for instance, IL-4, IL-5 and IL-10 (Bucy et al., 1995; Kelso, 1995; Mosmann and Sad, 1996). Whatever the behaviour of lymphocytes with respect to co-ordinate expression of cytokines at the single cell level, there can be no doubt that in many instances the maturing immune response is characterized overall by a phenotype of selective cytokine production. It is this selectivity that in turn determines the relative levels of IL-4 and IFN-γ and the likelihood that an IgE antibody response will be elicited.

If a major influence on homocytotropic antibody production is the preferential development of a Th2-type immune response then it is necessary to consider those factors that influence Th cell selectivity. It is clear that the nature of the immunogen itself and the characteristics of antigen processing and presenting cells are important (Weaver et al., 1988; Chang et al., 1990; Gajewski et al., 1991; Enk and Katz, 1994; Kolbe et al., 1994; Saito et al., 1994). Effective activation of T-lymphocytes requires two signals from antigen-presenting cells. Antigen itself provides the specificity, but a costimulatory signal is necessary also. Of the costimulatory molecules expressed by antigen-presenting cells the family of B7 proteins appear to be the most potent. This family comprises at least two molecules, B7-1 (CD80) and B7-2 (CD86), both of which interact with the shared receptors CD28/CTLA-4 on T-lymphocytes. It has been demonstrated recently that costimulation provided by B7-1 during antigen presentation favours Th1 cell development, whereas B7-2 favours Th2-type responses (Kuchroo et al., 1995). At the molecular level two other aspects of initial T-lymphocyte priming appear to be of particular relevance in the subsequent selectivity of Th cell responses. $CD4^+$ T-lymphocytes recognize foreign peptide or peptide–hapten conjugates in the context of MHC (major histocompatibility complex) class II (Ia) antigens expressed on the surface of antigen-presenting cells. It would appear that the nature of the foreign antigen–Ia complex is a major determinant of Th cell selectivity. In recent investigations it was found by Pfeiffer et al. (1995) that varying either the antigenic peptide or the Ia molecule could determine whether Th1- or Th2-type responses were generated. Those peptide–Ia complexes that interacted strongly with the T-cell receptor (TCR) for antigen favoured the generation of Th1 cells, while weaker binding with the TCR favoured priming of Th2 cells (Pfeiffer et al., 1995). Similar experiments were performed by Kumar et al. (1995) in which a battery of TCR ligands was used that differed with respect to binding affinity for Ia, but had unchanged T-lymphocyte specificity. Higher affinity ligands stimulated IFN-γ production, while ligands that displayed lower Ia binding affinity induced only IL-4 secretion (Kumar et al., 1995). Collectively these data demonstrate that the strength of the association between the immunogenic ligand and Ia and the affinity of interaction between the Ia–ligand complex and the TCR both influence Th cell selectivity. Superimposed upon this is the fact that high antigen dosage favours the differentiation of Th1 cells, while exposure of cells bearing the same TCR to lower concentrations of the same peptide antigen favours Th2 cell development (Constant et al., 1995).

In addition to the conditions under which T-lymphocytes recognize the priming antigen, the characteristics of Th-cell development are influenced by the availability of cytokines in the local microenvironment (Coffman et al., 1991; Swain et al., 1991; Abehsira-Amar et al., 1992; Hsieh et al., 1992). This represents arguably the most important determinant of Th-cell selectivity. Cytokines of Th1-type drive the development of Th1-cell responses, whereas IL-4, IL-10 and other Th2-cell products favour Th2-cell responses. Indeed, Th2-type responses fail to develop in IL-4 gene knockout mice (Kopf et al., 1993). These same cytokines are also selectively antagonistic insofar as IFN-γ inhibits the proliferation of Th2 cells and IL-10 inhibits cytokine synthesis by Th1 cells. Possibly the most important regulatory cytokines in the context of initial Th-cell phenotype determination are IL-10 and interleukin 12 (IL-12). As indicated above IL-10 is known to inhibit the ability of dendritic cells to induce the production of IFN-γ by T-lymphocytes (Macatonia et al., 1993). The major driver of Th1-cell development is IL-12, a heterodimeric cytokine produced primarily by phagocytic and dendritic cells (Trinchieri and Scott, 1994). It has been established that IL-12 will both promote the successful development of Th1 cells and inhibit Th2-type responses (Hsieh et al., 1993; McKnight et al., 1994; Reiner and Seder, 1995). The inhibition by IL-12 of Th2-cell function may be dependent upon its ability to stimulate the production of IFN-γ by natural killer (NK) cells. Conversely, the promotion of $CD4^+$ Th1-cell responses may operate through both IFN-γ-dependent and -independent mechanisms (Seder et al., 1993; Morris et al., 1994; Schmitt et al., 1994; Reiner and Seder, 1995; Wenner et al., 1996). Irrespective of the detailed mechanisms involved, IL-12 appears to influence markedly the manifestations of selective Th-cell responses in mice. Treatment with IL-12 inhibits Th2-cytokine expression and antigen-induced airway hyperresponsiveness in allergen-challenged mice and reduces significantly the number of inflammatory cells found in bronchoalveolar lavage fluid (Gavett et al., 1995). Conversely, neutralization of IL-12 in mice inhibits contact hypersensitivity, a cutaneous allergic response associated with Th1 cells and IFN-γ (Riemann et al., 1996).

If cytokines are themselves instrumental in regulating the development of differentiated Th-cell populations, then an intriguing question arises as to the provenance of these cytokines before a mature immune response has evolved. In the case of Th1-cell development then presumably the stimulation by dendritic cell- and/or macrophage-derived IL-12 of IFN-γ production by NK cells and the concerted action of IL-12 and IFN-γ on Th1-cell precursors may provide one answer. One suggestion regarding the evolution of Th2-cell responses has been that mast cells represent a source of IL-4 necessary for their selective differentiation (Romagnani, 1992). However, more recent investigations suggest that either NK1.1$^+$, CD4$^+$ T cells that express a limited repertoire of $\alpha\beta$-T-cell receptors and/or $\gamma\delta$-lymphocytes may provide a source of IL-4 rapidly following immune priming (Ferrick et al., 1995; Yoshimoto et al., 1995a, 1995b). It is possible also that other cells, in this instance as yet unidentified non-B/non-T cells, may represent an important source of IL-4 (and IL-6) to maintain and perpetuate Th2 responses (Aoki et al., 1995) (Figure 5.1).

The differential stimulation of Th-cell subpopulations has implications for the development of allergic reactions other than the integrity of IgE and IgG4 antibody responses. Th2 cytokines can influence the differentiation, distribution and function of cells involved in the elicitation of allergic hypersensitivity reactions. Interleukins 3, 4 and 10 are mast cell growth factors or cofactors (Smith and Rennick, 1986; Thompson-Snipes et al., 1991). In addition, IL-4 is capable of augmenting the secretory potential of mast cells and will enhance the release of serotonin following stimulation; an effect that is antagonized by IFN-γ (Coleman et al., 1991, 1992, 1993; Holliday et al., 1994). Interleukin 5 is a growth and differentiation factor for eosinophils (Yokota et al., 1987) and, as will be discussed later, regulates positively the accumulation of eosinophils at sites of allergen-induced respiratory reactions.

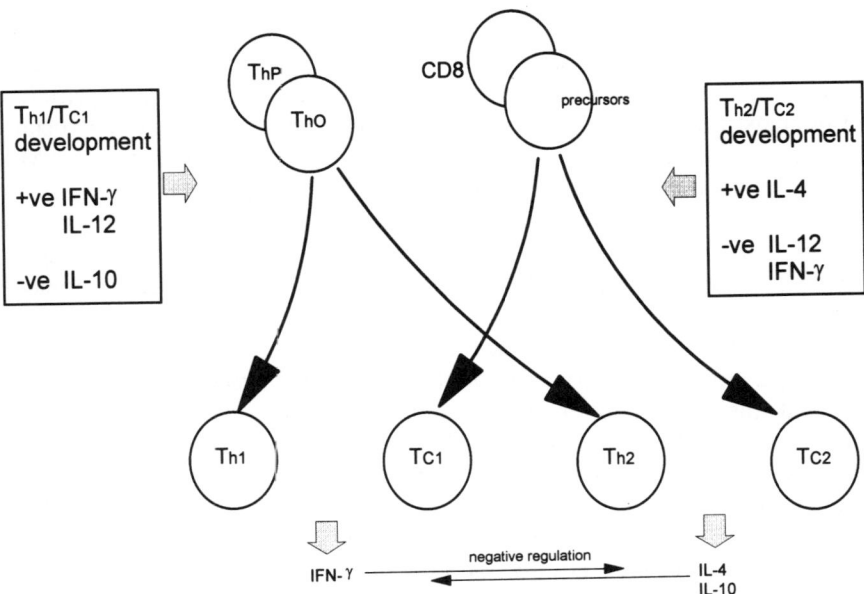

Figure 5.1 The influence of cytokines on the development of divergent CD4$^+$ (Th1 and Th2) and CD8$^+$ (Tc1 and Tc2) T-lymphocyte responses.

From the information available it can be concluded that for a chemical allergen to provoke the quality of immune response necessary for the induction of homocytotropic antibody and to facilitate sensitization of the respiratory tract, then selective Th2-type cell activation must be achieved. This, in turn, will depend upon the characteristics of the allergen itself, the nature and duration of exposure, the way in which it is presented to responsive T-lymphocytes and the availability of cytokines in the microenvironment. Evidence is available that those chemical allergens that are known in humans to induce sensitization of the respiratory tract and occupational asthma provoke in mice preferential Th2-type responses. In contrast, chemicals that are considered not to cause respiratory allergy in man, but which nevertheless are able to induce contact sensitization, stimulate selective Th1-type immune responses in mice (Dearman et al., 1992a; Kimber and Dearman, 1992, 1994). It has been demonstrated that epicutaneous exposure of mice to respiratory allergens such as trimellitic anhydride (TMA), phthalic anhydride and diphenylmethane diisocyanate results in the appearance of hapten-specific IgE antibodies and an increase in the serum concentration of IgE. Treatment of mice under identical conditions with concentrations of contact allergens (such as 2,4-dinitrochlorobenzene (DNCB) and oxazolone) that resulted in comparable T-lymphocyte proliferative responses, failed to stimulate IgE responses. It was observed also that contact allergens provoked a stronger IgG2a anti-hapten antibody response than did respiratory allergens; the IgG2a isotype in mice being known to be provoked by IFN-γ (Dearman and Kimber, 1991, 1992; Dearman et al., 1992c). It is possible to show that mice sensitized to TMA will mount both immediate- and delayed-type hypersensitivity reactions following dermal challenge with TMA. That this immediate oedematous response is a reflection of specific IgE production and the sensitization of dermal mast cells is supported by the fact that exposure to TMA is associated with the specific sensitization of mast cells in situ as determined by the TMA–protein conjugate induced release of serotonin in vitro (Dearman et al., 1992b; Holliday et al., 1993). Conversely, treatment of mice in the same way with DNCB failed to result in mast cell sensitization and dermal challenge of DNCB-sensitized animals resulted in delayed, but not immediate, cutaneous hypersensitivity reactions (Dearman et al., 1992b; Holliday et al., 1993). Following inhalation exposure of mice also TMA, but not DNCB, was found to induce IgE antibody (Dearman et al., 1991). The thesis that different classes of chemical allergen induce in mice qualitatively divergent immune responses consistent with the stimulation of selective T-cell activation is supported by analyses of cytokine production. Early following first exposure of mice, contact and respiratory chemical allergens provoke the production by draining lymph node cells of comparable cytokine profiles. However, following more chronic exposure, selective patterns of cytokine secretion are observed. Experience to date indicates that treatment of mice with chemical respiratory allergens results in the emergence of Th2-type cytokine production with comparatively high levels of IL-4 and IL-10 being secreted by activated lymph node cells, but only low levels of IFN-γ. The converse pattern is seen with contact allergens where vigorous IFN-γ production is induced, with relatively weak IL-4 and IL-10 responses (Dearman *et al.*, 1994, 1995, 1996a). These data indicate that in mice at least, chemicals known to cause respiratory sensitization in man provoke Th2-type responses (Figure 5.2).

It is important to emphasize here that functional subpopulations of T-lymphocytes are not restricted to the mouse. In man also there is evidence for a heterogeneity among Th cells and for the existence of Th1- and Th2-type populations (Romagnani, 1991; Romagnani *et al.*, 1992). Allergic respiratory hypersensitivity reactions and atopic asthma are associated with Th2-type cells. Biopsies of nasal turbinates from pollen-sensitive patients were examined following local provocation with allergen. Using *in situ* hybridization it was found that allergen challenge was associated with increased numbers of cells expressing mRNA for IL-3, IL-4 and IL-5 and that this correlated positively with tissue eosinophilia (Durham *et al.*, 1992). In subsequent analyses the majority of cells expressing IL-4 mRNA in the nasal mucosa following allergen challenge was characterized as being $CD3^+$ T-lymphocytes (Ying *et al.*, 1994). A predominance of Th2-type cells associated with eosinophil recruitment has been demonstrated also in bronchoalveolar lavage fluid prepared from patients with atopic asthma (Robinson *et al.*, 1992, 1993). Although a similar phenotypic analysis of lymphoid cells within the respiratory tract of patients with chemically-induced allergic hypersensitivity has not yet been performed, one study of isocyanate asthma is of interest. It was observed that isocyanate-induced occupational asthma and other forms of extrinsic asthma were characterized by very similar patterns of inflammatory cell infiltrates with the presence of activated T-lymphocytes and the recruitment of eosinophils

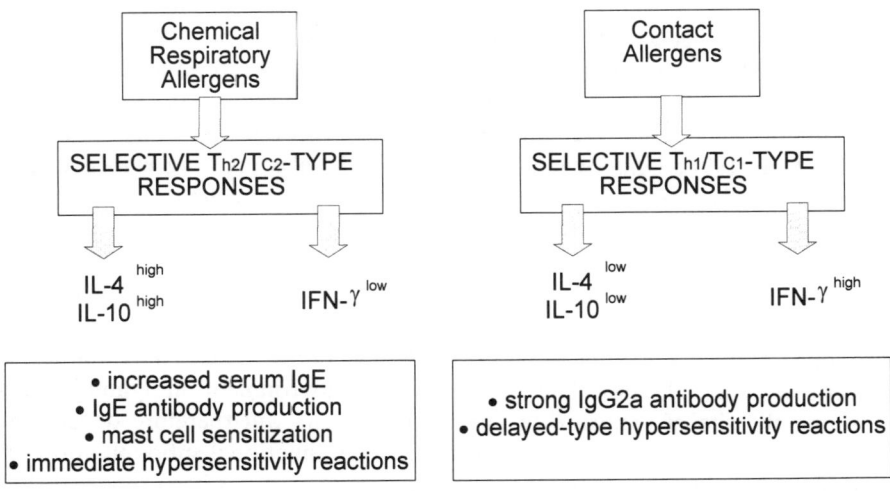

Figure 5.2 The characteristics of divergent immune responses induced in mice by chemical respiratory and contact allergens.

(Bentley et al., 1992). The conclusion drawn was that extrinsic asthma of diverse aetiology with respect to the inducing allergen involves similar immune and inflammatory reactions. It is also the case that atopic dermatitis lesional skin is associated with a high frequency of IL-4-producing Th2-type cells (van der Heijden et al., 1991). Similarly, clones of T-lymphocytes specific for house dust mite or pollen aeroallergens have been shown to produce Th2 cytokines, but not IFN-γ (Parronchi et al., 1991). A different picture is seen in allergic contact dermatitis where most, but not all, evidence points to a predominance of allergen-specific Th1-type cells. The majority of T-cell clones generated from the peripheral blood of patients with nickel sensitivity was shown to produce high levels of IFN-γ, but only low or undetectable levels of IL-4 and IL-5 (Kapsenberg et al., 1991, 1992).

Although most attention in recent years has focused on subpopulations of Th cells, there is the realization now that a similar, but not identical, functional heterogeneity is found among $CD8^+$ lymphocytes (Kemeny et al., 1994; Le Gros and Erard, 1994; Seder and Le Gros, 1995; Mosmann and Sad, 1996). Two populations, designated Tc1 and Tc2, have been described that display cytokine selectivity comparable, respectively, with Th1 and Th2 cells (Croft et al., 1994; Sad et al., 1995). In common with Th-cell subsets, it appears that differentiated Tc populations develop in response to similar cytokine signals. Thus, IL-12 and IFN-γ encourage the differentiation of precursors into Th1 or Tc1 cells, whereas IL-4 stimulates the development of Th2 or Tc2 cells (Croft et al., 1994; Noble and Kemeny, 1995; Noble et al., 1995; Sad et al., 1995; Mosmann and Sad, 1996). There is now considerable interest in the role of MHC class I-restricted CD8 subpopulations in immune regulation, the control of IgE antibody responses and in the elicitation of allergic hypersensitivity reactions. With regard to the latter there are several indications that $CD8^+$ cells may play an important role in contact hypersensitivity responses (Gocinski and Tigelaar, 1990; Fehr et al., 1994; Bour et al., 1995; Xu et al., 1996); the effector population being presumably IFN-γ-producing Tc1-type cells (Xu et al., 1996). In addition, it has been shown that $CD8^+$ cells of Tc1-type inhibit allergen-induced airways hyperresponsiveness in mice and rats (Renz et al., 1994; Underwood et al., 1995). There is growing evidence that these same cells play a role of considerable importance in the negative regulation of IgE antibody responses (Kemeny et al., 1992; Kemeny and Diaz-Sanchez, 1993; McMenamin and Holt, 1993; Renz et al., 1994; Underwood et al., 1995). Of potential relevance to the induction of chemical respiratory allergy are recent investigations which indicate that chemical sensitizers provoke in mice the appearance of IFN-γ-secreting $CD8^+$ T-lymphocytes (Dearman et al., 1996b). The cytokine microenvironment in lymph nodes draining the site of exposure to chemical allergens will therefore depend, at any one time, upon the balance achieved between functional subpopulations of both $CD4^+$ and $CD8^+$ T-lymphocytes. Very recently it has been reported that the capacity of $CD8^+$ cells to down-regulate IgE antibody responses to inhaled antigens is largely or wholly attributable to a small subset of extremely active antigen-responsive CD8$^+$ lymphocytes that produce high levels of IFN-γ and which display $\gamma\delta$-T-cell receptors (McMenamin et al., 1994, 1995). Transfer to syngeneic naive recipients of as few as 1000 CD8$^+$ $\gamma\delta^+$-T-cells from Brown Norway rats repeatedly exposed to antigen by inhalation was found to cause antigen-specific and IgE isotype-specific tolerance (McMenamin et al., 1995). The role of such cells in the regulation of IgE responses to chemical allergens has not yet been elucidated.

In summary, the quality of immune response that is provoked by antigens, including chemical allergens, is directed by the relative contribution of functional subpopulations of CD4$^+$ and CD8$^+$ T-lymphocytes and their soluble cytokine products. Those responses in which the development of Th2 and/or Tc2 cells predominate will favour the production of homocytotropic antibodies and will facilitate the elicitation of respiratory allergic reactions. Before leaving the induction of immune responses to chemical respiratory allergens it is instructive to consider the nature of the antigen that is recognized by responsive T-lymphocytes and the cells on which it may be presented.

5.4 Antigen and Antigen Presentation

In considering the nature and location of antigen-presenting cells that may be relevant for the induction of respiratory sensitization it is necessary first to identify the routes of exposure through which such sensitization may be achieved. It is assumed frequently that sensitization of the respiratory tract to chemical allergens is associated exclusively with inhalation exposure. Although there is no doubt that, in practice, respiratory sensitization to chemicals results usually from inhalation of the inducing allergen by a susceptible individual, this is not necessarily always the case. There is no reason to suppose that the quality and quantity of immune response necessary for effective sensitization of the respiratory tract could not result from dermal exposure; in the occupational setting as a consequence of accidents, spillages or splashes. There is experimental evidence and anecdotal clinical reports that indicate respiratory sensitization may result from dermal contact with the inducing chemical allergen (Kimber, 1996b, 1996c). It is no real surprise that routes of exposure other than by inhalation should induce sensitization of the respiratory tract to chemical allergens. Indeed, in some instances inhalation exposure may cause down-regulation of immune function, and in particular IgE responses (Fox and Siraganian, 1981; Holt et al., 1981; Doe et al., 1982; Sedgwick and Holt, 1985; Dearman and Botham, 1990). The induction of primary immune responses to contact allergens encountered on the skin and the development of skin sensitization is effected by epidermal Langerhans cells and possibly other cutaneous dendritic cells (DC) (Kimber and Cumberbatch, 1992). There is no reason presently to believe that these same populations of DC do not also participate in the stimulation of immune responses to respiratory chemical allergens encountered via skin surfaces. Presumably the network of DC identified within the airway epithelium will fulfil this role within the respiratory tract (Sertl *et al.*, 1986; Holt *et al.*, 1989, 1990a, 1990b; Pollard and Lipscomb, 1990; Schon-Hegrad *et al.*, 1991; Lipscomb *et al.*, 1995); DC being regarded as the mandatory antigen-presenting cells for the effective activation of virgin (unprimed) T-lymphocytes (Inaba and Steinman, 1984). Certainly there is evidence that acute inflammatory reactions in the respiratory tract and the induction of pulmonary immune responses are associated with an accumulation of DC in the airway epithelium and their subsequent migration to draining lymph nodes (McWilliam *et al.*, 1994; Xia *et al.*, 1995). It is likely that such DC in the respiratory tract are subject to local regulatory influences. The available evidence suggests that pulmonary alveolar macrophages function poorly as antigen-presenting cells and may in fact suppress T-lymphocyte responses (Holt, 1978, 1985). As a consequence it has been found that the elimination *in vivo* of alveolar macrophages is associated with a potentiation of pulmonary immune function (Thepen *et al.*, 1989, 1991). Of direct relevance to the induction of respiratory sensitization is the observation that removal of alveolar macrophages enhances selectively the vigour of IgE antibody responses to inhaled allergen (Thepen *et al.*, 1992). An important way in which pulmonary immune function may be down-regulated by alveolar macrophages is via their direct inhibition of the antigen-presenting cell properties of local DC (Holt *et al.*, 1985, 1988). It is believed therefore that the effectiveness of antigen presentation to T-lymphocytes in the respiratory tract is a function of a dynamic equilibrium between the potentiating properties of airways DC and the immunosuppressive characteristics of alveolar macrophages (Holt, 1993). It is likely as a result that the induction of respiratory sensitization to inhaled allergens will be influenced markedly by factors that perturb this equilibrium.

The exact form in which chemical allergens are displayed to responsive T-lymphocytes is uncertain. Chemical sensitizers are haptens, or (in the case of chemicals that require metabolic conversion to a protein-reactive species) prohaptens. As such they are non-immunogenic in their native state and must form stable associations with macromolecules (carriers) in order to stimulate an immune response. T-lymphocytes recognize peptides displayed in association with MHC class I ($CD8^+$ cells) or MHC class II ($CD4^+$ cells) membrane determinants. It appears that responsive T-lymphocytes recognize hapten, derived from the inducing chemical allergen, associated with peptides anchored to MHC molecules (Martin *et al.*, 1993; Martin and Weltzien, 1994; Cavani *et al.*, 1995; Kohler *et al.*, 1995; Weltzien *et al.*, 1996). What is not clear, however, is at what stage interaction of the chemical with protein occurs. In theory a protein-reactive chemical could interact directly with MHC-bound peptide. Alternatively, dendritic antigen-presenting cells might internalize, process and present extracellular protein that has interacted with chemical. What happens in practice is not known and it is possible also that covalently haptenized MHC molecules could represent antigenic epitopes for T-lymphocytes. Although most studies have focused primarily on the context in which skin sensitizing haptens are recognized, there is every reason to believe that chemical respiratory allergens will be handled and recognized by the immune system in a comparable manner.

Irrespective of the way in which allergen is processed and presented, and of the characteristics of immune responses that are induced as a consequence, the end result of relevance here is effective sensitization of the respiratory tract. The sensitized individual will mount more aggressive and accelerated secondary immune responses following subsequent exposure, resulting in local inflammatory reactions recognized clinically as respiratory hypersensitivity. The nature of the elicitation phase of chemical respiratory allergy is considered below.

5.5 T-Lymphocytes, Eosinophils and Respiratory Hypersensitivity Reactions

The symptoms associated with allergic hypersensitivity of the respiratory tract induced by chemicals are various and may include immediate-onset, late-phase or so called dual reactions (Salvaggio et al., 1988; Karol, 1992; Grammer, 1993). Immediate-type reactions are considered usually to be induced by degranulation of local mast cells, mediated by IgE or other homocytotropic antibody, and the release of inflammatory mediators, including vasoactive amines and leukotrienes, that initiate vasodilation, the contraction of smooth muscle and bronchoconstriction. Late asthmatic responses to chemical allergens are manifest 2 to 8 hours following exposure and are characterized by the infiltration of mononuclear cells and an increased number of leucocytes in bronchoalveolar lavage fluid. Chronic inflammation plays a critical role in asthma and, in addition to the accumulation in bronchial mucosa of leucocytes, is associated with the destruction and sloughing of airway epithelial cells, the production of mucus and subepithelial fibrosis secondary to collagen deposition (Beasley et al., 1989; Roche et al., 1989). Central to the development of chronic bronchial inflammation and injury are eosinophils acting in concert with T-lymphocytes (Durham and Kay, 1985; Gleich, 1990; Barnes and Lee, 1992; Bentley et al., 1992; Corrigan, 1992; Corrigan and Kay, 1992; Seminario and Gleich, 1994). The exact role of eosinophils in allergic hypersensitivity reactions and asthma is presently unclear, although when activated these cells are clearly able to induce tissue injury by the release of preformed toxic proteins and by the *de novo* synthesis of oxygen radicals. Activation of eosinophils results in the release of granule proteins including major basic protein (MBP), eosinophil peroxidase, eosinophil cationic protein and eosinophil-derived neurotoxin (Gleich and Adolphson, 1986; Weller, 1991). These factors act as cationic toxins and cytostimulants and their levels correlate closely with the severity of asthmatic symptoms (Frigas et al., 1981; Filley et al., 1982). Moreover, the direct administration of MBP into the trachea of primates causes a dose-dependent increase in airway hyperreactivity (Gundel et al., 1991). The directed movement of eosinophils, their accumulation and activation are regulated by cytokines, chemokines and other soluble factors (Robinson, 1993; Bacon and Schall, 1996; Humbert, 1996). Of particular relevance appears to be IL-5, a product of Th2 lymphocytes, that promotes eosinophil production in, and release from, the bone marrow and the movement, activation and survival of these cells (Yokota et al., 1987; Lopez et al., 1988; Yamaguchi et al., 1988; Wang et al., 1989; Gulbenkian et al., 1992). Mice expressing a transgene for IL-5 display marked eosinophilia in bronchoalveolar lavage fluid and exhibit more profound airway hyperreactivity in response to allergen challenge (Iwamoto and Takatsu, 1995). A similar increase in airway hyperresponsiveness was observed in guinea pigs that had been treated, by intratracheal injection, with heterologous recombinant IL-5 (Iwami et al., 1993). Conversely, it has been demonstrated that sensitized mice which lack IL-5 fail to display the eosinophilia, airways hyperreactivity and lung damage that normally result from inhalation challenge with the relevant aeroallergen (Foster et al., 1996). It should be mentioned, however, that there may not exist a mandatory role for IL-5 and eosinophilia in the effective elicitation of all forms of allergic respiratory hypersensitivity. It has been shown recently in another mouse model that treatment with a neutralizing anti-IL-5 antibody inhibited markedly eosinophilia, but failed to prevent the development of airways hyperresponsiveness following appropriate sensitization and challenge (Corry et al., 1996). In the latter investigation airways hyperresponsiveness was nevertheless inhibited by the depletion of $CD4^+$ T-lymphocytes or by the neutralization of IL-4 during sensitization (Corry et al., 1996). It is possible that these apparently conflicting data point to the existence of two (or more) independent, but overlapping, mechanisms that can result in airway responses in the sensitized lung (Drazen et al., 1996). Other evidence also points to a key role for IL-4 in allergic hypersensitivity reactions in the respiratory tract (Brusselle et al., 1994; Lukacs et al., 1994; Renz et al., 1996).

The development of eosinophilia is antagonized by IFN-γ, possibly acting via the inhibition of $CD4^+$ T-lymphocyte infiltration (Iwamoto et al., 1993; Kung et al., 1995). These data again reflect the reciprocal effects of Th1 and Th2 cells and their soluble cytokine products. There is good evidence that in many instances the elicitation of allergic hypersensitivity reactions in the respiratory tract is favoured by, or dependent upon, the infiltration and activation of Th2-type cells that are a source of the cytokines necessary for pulmonary inflammation (Ricci et al., 1993). To what extent the accumulation of these cells at sites of allergen challenge in the respiratory tract is facilitated by IgE-dependent mast cell degranulation and vasodilation is not clear.

Although complementary mechanisms may exist, it is apparent that the accumulation of eosinophils is an important event in the development of chronic bronchial inflammation and asthma. While IL-5 is undoubtedly a key cytokine in this process, others such as IL-3 and GM-CSF may also play a role (Humbert, 1996). For effective migration from the blood to the bronchial mucosa eosinophils must adhere to the vascular endothelium, extracellular matrices and tissue cells. Such interactions are mediated through the recognition of relevant adhesion molecules and several of these have been implicated in the accumulation of eosinophils. One such is vascular cell adhesion molecule 1 (VCAM-1), which has as its integrin ligand very late antigen 4 (VLA-4). It has been shown that the expression of VCAM-1, and of other adhesion molecules, is upregulated rapidly in the bronchial mucosa following allergen challenge of atopic asthmatics (Bentley et al., 1993). As eosinophils express VLA-4 and since VCAM-1 expression can be augmented by cytokines, including IL-4, it has been proposed that the selective recruitment of eosinophils into the bronchial mucosa is effected, at least in part, by the IL-4-mediated upregulation of VCAM-1 (Schleimer et al., 1992; Seminario and Gleich, 1994; Humbert, 1996). If such a mechanism is important then it might go some way to explain the need in some models for IL-4 in the development of airways hyperreactivity.

The directed recruitment of eosinophils into the bronchial mucosa is an area of some interest, and a number of chemokines have been found to influence eosinophil trafficking (Resnick and Weller, 1993; Baggiolini and Dahinden, 1994; Baggiolini et al., 1994; Bacon and Schall, 1996; Frew, 1996; Humbert, 1996). However, a novel chemokine, eotaxin, has been described recently that appears to be unique among such mediators insofar as it causes the selective infiltration of eosinophils into the lungs and skin (Griffiths-Johnson et al., 1993; Jose et al., 1994). In sensitized guinea pigs eotaxin mRNA expression in the lung is enhanced following inhalation challenge (Rothenberg et al., 1995). It has been proposed that this chemokine and IL-5 act together to induce eosinophil accumulation, with eotaxin (and other chemokines) acting locally to encourage the movement of eosinophils from the blood and IL-5 acting hormonally to release bone marrow eosinophils into the circulation (Collins et al., 1995).

Like the induction phase of sensitization, the elicitation of respiratory allergic reactions is dependent upon the co-ordinated interaction between immunologically competent cells and cytokines. Very complex immune mechanisms are involved that if activated correctly and with sufficient vigour result in the development of respiratory allergy. It is apparent, however, that not all individuals are susceptible to chemical-induced respiratory hypersensitivity. With respect to common protein aeroallergens, atopy (a genetic predisposition toward the production of IgE) is an important risk factor. Chemical respiratory allergy appears not to be more common among atopics and the inherent factors that govern inter-individual variations in the likelihood of sensitization are unknown. It may be, however, that external factors play their part in susceptibility to chemical allergy.

5.6 Environmental Factors, Respiratory Sensitization and Asthma

In many countries the prevalence of asthma is rising; an increase that is apparently independent of improved diagnosis or altered diagnostic criteria

(Cullinan, 1988; Gergen et al., 1988; Burney et al., 1990; Hahhtela et al., 1990; Weitzman et al., 1992; Aberg et al., 1995). This increased prevalence has been too rapid to be a function of changes in the gene pool and there has consequently been a growing interest in the possibility that environmental factors and, in particular, environmental pollution, play an important role in the development and/or severity of asthma (Wardlaw, 1993; Barnes, 1994; Newman Taylor, 1995).

In theory there are a number of ways in which environmental air pollutants may influence the pathophysiology of asthma. They might themselves induce or augment airway inflammation or act as inciters of reactions in airways that are already hyperresponsive. Alternatively, it may be that some pollutants have the ability to enhance or modify immune responses to inhaled antigens such as to influence the effectiveness of sensitization and/or the vigour of elicitation reactions. While there is little or no direct evidence to support the hypothesis that the increased prevalence of asthma is due in part to the impact of environmental pollutants on pulmonary sensitization, it is instructive to review the experimental data available and to speculate on possible mechanisms. In this context several processes may be relevant, although it must be emphasized that they need not operate independently of each other. It is possible that increased immunogenicity of inhaled antigen might result from an association with particulate pollutants that act as adjuvants. Alternatively, immune responsiveness could be enhanced by more effective handling, processing and presentation of inhaled antigen in the respiratory tract or from the elimination by pollutants of cells that negatively regulate pulmonary immune function. With respect to effects on suppressive or regulatory cells, it can be envisaged that factors which influence the balance between dendritic cells and alveolar macrophages in the respiratory tract (Holt, 1993) could impact upon the effectiveness of local antigen presentation.

It is known that the migration, maturation and function of epidermal Langerhans cells (LC) are regulated by cytokines produced locally (Kimber and Cumberbatch, 1992). It is assumed that the network of DC found within the airways epithelium will be subject to the same or similar regulatory influences of cytokines. It is known that the respiratory epithelium is a source of cytokines, including those that have been found previously to influence the behaviour of LC (Churchill et al., 1992; Cromwell et al., 1992). Studies performed in vitro with human bronchial epithelial cells have revealed that exposure to comparatively low concentrations of nitrogen dioxide results in cell damage, reduced ciliary beat frequency and the release of several cytokines including IL-8, GM-CSF and tumour necrosis factor-α (TNF-α) (Devalia et al., 1993a, 1993b). Whether the release of such cytokines will serve to enhance the migration of allergen-bearing pulmonary DC to draining lymph nodes and/or augment their ability to present allergen to responsive T-lymphocytes is not known. It has been demonstrated recently that exposure of mice to atmospheres of formaldehyde results in the stimulation of enhanced IgE antibody responses to ovalbumin following intranasal administration of the protein allergen (Tarkowski and Gorski, 1995). It is possible that such augmentation of IgE responses results from the influence of formaldehyde on the airway epithelium and/or alveolar macrophages in such a way as to promote the effective presentation of allergen.

Another mechanism through which environmental pollutants may influence pulmonary immune responses to inhaled allergens is via a direct effect on the vigour of relevant antibody responses. There is increasing evidence that some pollutants and tobacco smoke act as adjuvants for IgE production. Several investigations have indicated that tobacco smoking is associated with an increased concentration of IgE in the plasma (Gerrard et al., 1980; Burrows et al., 1981; Zetterstrom et al., 1981; Omenaas et al., 1994). In addition, the frequency of IgE sensitization to chemical and protein respiratory allergens is higher in those who smoke (Anon, 1985; Venables et al., 1985). The incidence of allergic disease is increased in the offspring of mothers who smoke cigarettes and maternal smoking is associated also with higher levels of cord blood IgE (Magnusson, 1986).

Experimental studies in mice indicate that diesel exhaust particles (DEP) enhance the production of specific IgE antibody. The intraperitoneal injection

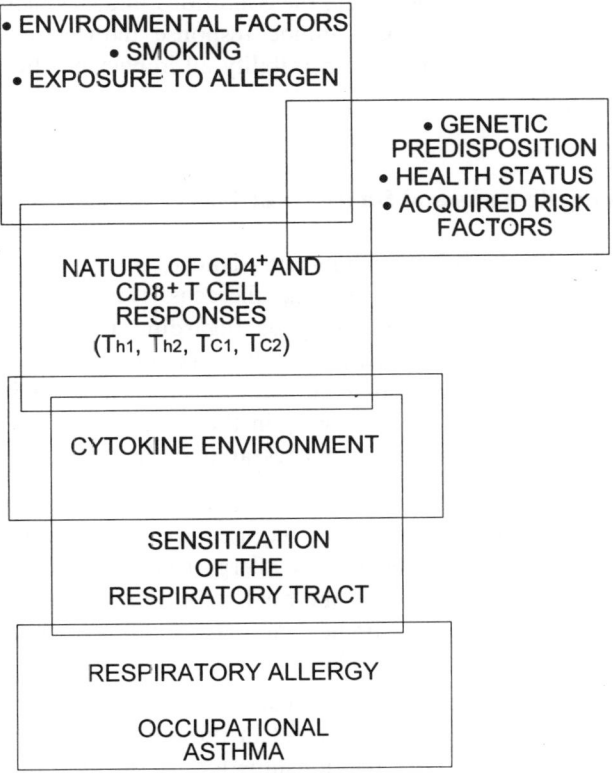

Figure 5.3 Determinants of the allergic response to chemical respiratory sensitizers.

of mice with a mixture of DEP and ovalbumin was shown to induce a more vigorous IgE antibody response to the protein allergen than that displayed by animals treated with ovalbumin alone (Muranka et al., 1986). Comparable results were obtained when Japanese cedar pollen was used in place of ovalbumin as the inducing antigen (Muranka et al., 1986). In subsequent investigations by the same group it was shown that DEP also enhanced IgE responses when administered intranasally (Takafuji et al., 1987). It is possible that the increased IgE responses observed following treatment with DEP are secondary to the stimulation of augmented IL-4 production (Fujimaki et al., 1994). Alternatively, DEP, or soluble extracts of DEP, may interact directly with activated B-lymphocytes to enhance the production of IgE (Diaz-Sanchez et al., 1994; Takenaka et al., 1995).

Although relevant immunobiological mechanisms can be envisaged, it must be emphasized that there are currently available no compelling data to indicate that normal exposure to environmental pollutants, other than perhaps tobacco smoke, will induce or enhance allergic sensitization. Nevertheless, the possibility that environmental factors influence susceptibility to chemical respiratory allergens can not be excluded.

A complex picture of respiratory sensitization to chemical allergens emerges where the nature of the allergen itself, the conditions under which it is encountered, the characteristics of the immune response induced and congenital and/or acquired determinants of susceptibility all impact on the likelihood of sensitization (Figure 5.3).

5.7 Concluding Remarks and Future Directions

In recent years there have been significant advances in the characterization of immune responses induced by chemical allergens. Despite such progress it must be acknowledged that a consensus regarding the immunobiological mechanisms that, in practice, result in respiratory sensitization has yet to be achieved. At the centre of the debate are questions regarding the importance of IgE and other homocytotropic antibody in the development of sensitization. This and other priority areas for further research are listed below.

1. The development and/or acceptance of experimental animal models for the elucidation of relevant mechanisms requires that the immunological basis or bases for occupational respiratory allergy to chemicals are understood. A major achievement would be the final resolution of the relative importance of homocytotropic antibodies and other immune mechanisms in respiratory sensitization to chemicals.

2. The physiochemical basis for chemical respiratory allergy remains to be elucidated. Not only would this assist in the development of appropriate models of structure-activity relationships for use in the predictive assessment of respiratory sensitization potential, but also provide important

clues as to the characteristics of chemical allergens that determine the nature of immune responses they will provoke.

3 Linked with the above is the need for a more sophisticated understanding of the way in which chemical allergens are handled and processed and presented to the immune system. It is probable that early events after initial encounter with the inducing chemical allergen (such as the stimulation of cytokine production) combined with the characteristics of relevant antigen-presenting cells and the context in which hapten is presented to responsive T-lymphocytes, determine both the quality and vigour of induced immune responses.

4 It remains to be established to what extent the route of exposure to chemical allergen impacts on the effectiveness of sensitization. Although there is reason to believe that sensitization of the respiratory tract can be achieved following dermal contact with an appropriate chemical allergen, it is not yet clear whether the characteristics of immune responses induced by chemical allergens in different tissue compartments vary in important respects.

5 The nature of the inflammatory processes that result in respiratory hypersensitivity reactions to chemical allergens, and the development of occupational asthma require further clarification. Of special interest are the relative contributions of particular cytokines, chemokines and other inflammatory mediators.

6 The factors that govern individual susceptibility to chemical respiratory allergy and the impact of environmental conditions on the effectiveness of sensitization remain unclear. Risk assessment and risk management of chemical respiratory allergy would benefit considerably from a clearer understanding of genetic predisposition and the influence of air quality on sensitization.

References

ABEHSIRA-AMAR, O., GILBERT, M., JOLIY, M., THEZE, J. & JANKOVIC, D. L. (1992) IL-4 plays a dominant role in the differential development of Th0 into Th1 and Th2 cells. *Journal of Immunology*, **148**, 3820–3829.

ABERG, N., HESSELMAR, B., ABERG, B. & ERIKSSON, B. (1995) Increase of asthma, allergic rhinitis and eczema in Swedish schoolchildren between 1979 and 1991. *Clinical and Experimental Allergy*, **25**, 815–819.

ANON. (1985) Smoking, occupation and allergic lung disease. *Lancet*, **i**, 965.

AOKI, I., KINZER, C., SHIRAI, A., PAUL, W. E. & KLINMAN, D. M. (1995) IgE receptor-positive non-B/non-T cells dominate the production of interleukin 4 and interleukin 6 in immunized mice. *Proceedings of the National Academy of Sciences USA*, **92**, 2534–2538.

ARMITAGE, R. J., MACDUFF, B. M., SPRIGGS, M. K. & FANSLOW, W. C. (1993) Human B cell proliferation and Ig secretion induced by recombinant CD40 ligand are modulated by soluble cytokines. *Journal of Immunology*, **150**, 3671–3679.

AVERSA, G., PUNNONEN, J., COCKS, B. G., DE WAAL MALEFYT, R., VEGA, F. JR., ZURAWSKI, S. M., ZURAWSKI, G. & DE VRIES, J. E. (1993) An interleukin 4 (IL-4) mutant protein inhibits both IL-4 or IL-13 induced human immunoglobulin G4 (IgG4) and IgE synthesis and B cell proliferation: support for a common component shared by IL-4 and IL-13 receptors. *Journal of Experimental Medicine*, **178**, 2213–2218.

AZUMA, M., HIRANO, T., MIYAJIMA, H., WATANABE, N., YAGITA, H., ENOMOTO, S., FURUSAWA, S., OVARY, Z., KINASHI, T. & HONJA, T. (1987) Regulation of IgE production in SJA/9 and nude mice. Potentiation of IgE production by recombinant interleukin 4. *Journal of Immunology*, **139**, 2538–2544.

BACON, K. B. & SCHALL, T. J. (1996) Chemokines as mediators of allergic inflammation. *International Archives of Allergy and Immunology*, **109**, 97–109.

BAGGIOLINI, M. & DAHINDEN, C. A. (1994) CC chemokines in allergic inflammation. *Immunology Today*, **104**, 27–29.

BAGGIOLINI, M., DEWALD, B. & MOSER, B. (1994) Interleukin-8 and related chemotactic cytokines – CXC and CC chemokines. *Advances in Immunology*, **55**, 97–179.

BARNES, P. J. (1994) Air pollution and asthma. *Postgraduate Medical Journal*, **70**, 319–325.

BARNES, P. J. & LEE, T. H. (1992) Recent advances in asthma. *Postgraduate Medical Journal*, **68**, 942–953.

BEASLEY, R., ROCHE, W. R., ROBERTS, J. A. & HOLGATE, S. T. (1989) Cellular events in the bronchi in mild asthma and after bronchial provocation. *American Review of Respiratory Disease*, **139**, 806–817.

BENDELAC, A. & SCHWARTZ, R. H. (1991) Th0 cells in the thymus. The question of T-helper lineages. *Immunological Reviews*, **123**, 169–188.

BENTLEY, A. M., MAESTRELLI, P., SAETTA, M., FABBRI, L. M., ROBINSON, D. S., BRADLEY, B. L., JEFFREY, P. K., DURHAM, S. R. & KAY, A. B. (1992) Activated T-lymphocytes and eosinophils in the bronchial mucosa in isocyanate-induced asthma. *Journal of Allergy and Clinical Immunology*, **89**, 821–829.

BENTLEY, A. M., DURHAM, S. R., ROBINSON, D. S., MENZ, G., STORZ, C., CROMWELL, O., KAY, A. B. & WARDLAW, A. J. (1993) Expression of endothelial and leukocyte adhesion molecules intercellular adhesion molecule-1, E-selectin and vascular adhesion molecule-1 in the bronchial mucosa in steady-state and allergen-induced asthma. *Journal of Allergy and Clinical Immunology*, **92**, 857–868.

BONNEFOY, J-Y., POCHON, S., AUBRY, J-P., GRABER, P., GAUCHAT, J-F., JANSEN, K. & FLORES-ROMO, L. (1993) A new pair of surface molecules involved in human IgE regulation. *Immunology Today*, **14**, 1–2.

BORGER, P., KAUFFMAN, H. F., POSTMA, D. S. & VELLENGA, E. (1996) IL-7 differentially modulates the expression of IFN-γ and IL-4 in activated human T lymphocytes by transcriptional and post-transcriptional mechanisms. *Journal of Immunology*, **156**, 1333–1338.

BOUR, H., PEYRON, E., GAUCHERAND, M., GARRIGUE, J-L., DESVIGNES, C., KAISERLIAN, D., REVILLARD, J.-P. & NICOLAS, J.-F. (1995) Major histocompatibility complex class 1-restricted CD8$^+$ T cells and class II-restricted CD4$^+$ T cells, respectively, mediate and regulate contact sensitivity to dinitrofluorobenzene. *European Journal of Immunology*, **25**, 3006–3010.

BRUSSELLE, G. G., KIPS, J. C., TAVERNIER, J. H., VAN DER HEYDEN, J. G., CUVELIER, C. A., PAUWELS, R. A. & BLUETHMAN, H. (1994) Attenuation of allergic airway inflammation in IL-4 deficient mice. *Clinical and Experimental Allergy*, **24**, 73–80.

BUCY, R. P., KARR, L., HUANG, G-Q., LI, J., CARTER, D., HONJO, K., LEMONS, J. A., MURPHY, K. M. & WEAVER, C. T. (1995) Single cell analysis of cytokine gene coexpression during CD4$^+$ T-cell phenotype development. *Proceedings of the National Academy of Sciences USA*, **92**, 7565–7569.

BURNEY, P. G. J., CHINN, S. & ROMA, R. J. (1990) Has the prevalence of asthma changed? Evidence from the national study of health and growth 1973–86. *British Medical Journal*, **300**, 1306–1310.

BURROWS, B., HALONEN, M., BARBEE, R. A. & LEBOWITZ, M. D. (1981) The relationship of serum immunoglobulin E to cigarette smoking. *American Review of Respiratory Disease*, **124**, 523–525.

BURSTEIN, H. J., TEPPER, R. I., LEDER, P. & ABBAS, A. K. (1991) Humoral immune functions in IL-4 transgenic mice. *Journal of Immunology*, **147**, 2950–2956.

CAVANI, A., HACKETT, C. J., WILSON, K. J., ROTHBARD, J. B. & KATZ, S. I. (1995) Characterization of epitopes recognized by hapten-specific CD4$^+$ T cells. *Journal of Immunology*, **154**, 1232–1238.

CHANG, T-L., SHEA, C. M., URIOSTE, S., THOMPSON, R. C., BOOM, W. H. & ABBAS, A.K. (1990) Heterogeneity of helper/inducer T lymphocytes. III. Responses of IL-2- and IL-4-producing (Th1 and Th2) clones to antigen presented by different accessory cells. *Journal of Immunology*, **145**, 2803–2808.

CHAN-YEUNG, M. (1990) Occupational asthma. *Chest*, **98**, 148S–161S.

(1995) Occupational asthma. *Environmental Health Perspectives*, **103**, 249–252.

CHRETIEN, I., PENE, J., BRIERE, F., DE WAAL MALEFYT, R., ROUSSET, F. & DE VRIES, J. (1990) Regulation of human IgE synthesis. 1. Human IgE synthesis *in vitro* is determined by the reciprocal antagonistic effects of interleukin 4 and interferon-γ. *European Journal of Immunology*, **20**, 243–251.

CHURCHILL, L., FRIEDMAN, B., SCHLEIMER, R. P. & PROUD, D. (1992) Production of granulocyte-macrophage colony-stimulating factor by cultured human tracheal epithelial cells. *Immunology*, **75**, 189–195.

COCKS, B. G., DE WAAL MALEFYT, R., GALIZZI, J-P., DE VRIES, J. E. & AVERSA, G. (1993) IL-13 induces proliferation and differentiation of human B cells activated by the CD40 ligand. *International Immunology*, **5**, 657–661.

COFFMAN, R. L., VARKILA, K., SCOTT, P. & CHATELAIN, R. (1991) Role of cytokines in the differentiation of CD4$^+$ T-cell subsets *in vivo*. *Immunological Reviews*, **123**, 189–207.

COLEMAN, J. W., BUCKLEY, M. G., HOLLIDAY, M. R. & MORRIS, A. G. (1991) Interferon-γ inhibits serotonin release from mouse peritoneal mast cells. *European Journal of Immunology*, **21**, 2559–2564.

COLEMAN, J. W., HOLLIDAY, M. R. & BUCKLEY, M. G. (1992) Regulation of the secretory function of mouse peritoneal mast cells by IL-3, IL-4 and IFN-γ. *International Archives of Allergy and Immunology*, **99**, 408–410.

COLEMAN, J. W., HOLLIDAY, M. R., KIMBER, I., ZSEBO, K. M. & GALLI, S. J. (1993) Regulation of mouse peritoneal mast cell secretory function by stem cell factor, IL-3 or IL-4. *Journal of Immunology*, **150**, 556–562.

COLLINS, P. D., MARLEAU, S., GRIFFITHS-JOHNSON, D. A., JOSE, P. & WILLIAMS, T. J. (1995) Cooperation between interleukin-5 and the chemokine eotaxin to induce eosinophil accumulation *in vivo*. *Journal of Experimental Medicine*, **182**, 1169–1174.

CONSTANT, S., PFEIFFER, C., WOODARD, A., PASQUALINI, T & BOTTOMLY, K. (1995) Extent of T cell receptor ligation can determine the functional differentiation of naive $CD4^+$ T cells. *Journal of Experimental Medicine*, **182**, 1591–1596.

CORRIGAN, C. J. (1992) Allergy of the respiratory tract. *Current Opinion in Immunology*, **4**, 798–804.

CORRIGAN, C. J. & KAY, A. B. (1992) T cells and eosinophils in the pathogenesis of asthma. *Immunology Today*, **13**, 501–507.

CORRY, D. B., FOLKESSON, H. G., WARNOCK, M. L., ERLE, D. J., MATTHAY, M. A., WIENER-KRONISH, J. P. & LOCKSLEY, R. M. (1996) Interleukin 4, but not interleukin 5 or eosinophilia, is required in a murine model of acute airway hyperreactivity. *Journal of Experimental Medicine*, **183**, 109–117.

CROFT, M., CARTER, L., SWAIN, S. L. & DUTTON, R. W. (1994) Generation of polarized antigen-specific CD8 effector populations: reciprocal action of interleukin (IL)-4 and IL-12 in promoting type 2 versus type 1 cytokine profiles. *Journal of Experimental Medicine*, **180**, 1715–1728.

CROMWELL, O., HAMID, Q., CORRIGAN, C. J., BARKANS, J., MENG, Q., COLLINS, P. D. & KAY, A. B. (1992) Expression and generation of interleukin-8, IL-6 and granulocyte-macrophage colony-stimulating factor by bronchial epithelial cells and enhancement by IL-1β and tumour necrosis factor-α. *Immunology*, **77**, 330–337.

CULLINAN, P. (1988) Respiratory disease in England and Wales. *Thorax*, **43**, 949–954.

DEARMAN, R. J. & BOTHAM, P. A. (1990) Inhalation exposure to respiratory sensitising chemicals down-regulates guinea pig IgE and pulmonary responses. *International Archives of Allergy and Applied Immunology*, **92**, 425–432.

DEARMAN, R. J. & KIMBER, I. (1991) Differential stimulation of immune function by respiratory and contact chemical allergens. *Immunology*, **72**, 563–570.

(1992) Divergent immune responses to respiratory and contact chemical allergens: antibody elicited by phthalic anhydride and oxazolone. *Clinical and Experimental Allergy*, **22**, 241–250.

DEARMAN, R. J., HEGARTY, J. M. & KIMBER, I. (1991) Inhalation exposure of mice to trimellitic anhydride induces both IgG and IgE anti-hapten antibody. *International Archives of Allergy and Applied Immunology*, **95**, 70–76.

DEARMAN, R. J., BASKETTER, D. A., COLEMAN, J. W. & KIMBER, I. (1992a) The cellular and molecular basis for divergent allergic responses to chemicals. *Chemical-Biological Interactions*, **84**, 1–10.

DEARMAN, R. J., MITCHELL, J. A., BASKETTER, D. A. & KIMBER, I. (1992b) Differential ability of occupational chemical contact and respiratory allergens to cause immediate and delayed dermal hypersensitivity reactions in mice. *International Archives of Allergy and Immunology*, **97**, 315–321.

DEARMAN, R. J., SPENCE, L. M. & KIMBER, I. (1992c) Characterization of murine immune responses to allergenic diisocyanates. *Toxicology and Applied Pharmacology*, **112**, 190–197.

DEARMAN, R. J., RAMDIN, L. S. P., BASKETTER, D. A. & KIMBER, I. (1994) Inducible IL-4-secreting cells provoked in mice during chemical sensitization. *Immunology*, **81**, 551–557.

DEARMAN, R. J., BASKETTER, D. A. & KIMBER, I. (1995) Differential cytokine production following chronic exposure of mice to chemical respiratory and contact allergens. *Immunology*, **86**, 545–550.

DEARMAN, R. J., BASKETTER, D. A. & KIMBER, I. (1996a) Characterization of chemical allergens as a function of divergent cytokine secretion profiles induced in mice. *Toxicology and Applied Pharmacology*, **138**, 308–316.

DEARMAN, R. J., MOUSSAVI, A., KEMENY, D. M. & KIMBER, I. (1996b) Contribution of $CD4^+$ and $CD8^+$ T lymphocyte subsets to the cytokine secretion patterns induced in mice during sensitization to contact and respiratory chemical allergens. *Immunology* (in press).

DEL PRETE, G. F., MAGGI, E., PARRONCHI, P., CHRETIEN, I., TIRI, A., MACCHIA, D., RICHI, M., BANCHEREAU, J., DE VRIES, J. & ROMAGNANI, S. (1988) IL-4 is an essential factor for the IgE synthesis induced *in vitro* by human T cell clones and their supernatants. *Journal of Immunology*, **140**, 4193–4198.

DEVALIA, J. L., CAMPBELL, A. M., SAPSFORD, R. J., RUSZNAK, C., QUINT, D., GODARD, P., BOUSQUET, J. & DAVIES, R. J. (1993a) Effect of nitrogen dioxide on synthesis of inflammatory cytokines expressed by human bronchial epithelial cells *in vitro*. *American Journal of Respiratory Cell and Molecular Biology*, **9**, 271–278.

DEVALIA, J. L., SAPSFORD, R. J., CUNDELL, D. R., RUSZNAK, C., CAMPBELL, A. M. & DAVIES, R. J. (1993b) Human bronchial epithelial cell dysfunction following exposure to nitrogen dioxide. *European Respiratory Journal*, **6**, 1308–1316.

DIAZ-SANCHEZ, D., DOTSON, A. R., TAKENAKA, H. & SAXON, A. (1994) Diesel exhaust particles induce local IgE production *in vivo* and alter the pattern of IgE messenger RNA isoforms. *Journal of Clinical Investigation*, **94**, 1417–1425,

DOE, J. E., HICKS, R. & MILBURN, G. M. (1982) Specific inhibition of contact sensitivity in the guinea pig following toluene diisocyanate inhalation. *International Archives of Allergy and Applied Immunology*, **68**, 275–279.

DRAZEN, J. M., ARM, J. P. & AUSTEN, K. F. (1996) Sorting out the cytokines of asthma. *Journal of Experimental Medicine*, **183**, 1–5.

DREXLER, H., SCHALLER, K.-H., WEBER, A., LETZEL, S. & LEHNERT, G. (1993) Skin prick tests with solutions of acid anhydrides in acetone. *International Archives of Allergy and Immunology*, **100**, 251–255.

DURHAM, S. R. & KAY, A. B. (1985) Eosinophils, bronchial hyperreactivity and late phase asthmatic reactions. *Clinical Allergy*, **15**, 411–418.

DURHAM, S. R., YING, S., VARNEY, V. A., JACOBSON, M. R., SUDDERICK, R. M., MACKAY, I. S., KAY, A. B. & HAMID, Q. A. (1992) Cytokine messenger RNA expression for IL-3, IL-4, IL-5 and granulocyte/macrophage colony-stimulating factor in the nasal mucosa after local allergen provocation: relationship to tissue eosinophilia. *Journal of Immunology*, **148**, 2390–2394.

ENK, A. H. & KATZ, S. I. (1994) Heat-stable antigen is an important costimulatory molecule on epidermal Langerhans cells. *Journal of Immunology*, **152**, 3264–3270.

FABBRI, L. M., MAESTRELLI, P., SAETTA, M. & MAPP, C. M. (1994) Mechanisms of occupational asthma. *Clinical and Experimental Allergy*, **24**, 628–635.

FEHR, B. S., TAKASHIMA, A., MATSUE, H., GEROMETTA, J. S., BERGSTRESSER, P. R. & CRUZ, P. D., JR. (1994) Contact sensitization induces proliferation of heterogeneous populations of hapten-specific T cells. *Experimental Dermatology*, **3**, 189–197.

FERRICK, D. A., SCHRENZEL, M. D., MULVANIA, T., HSIEH, B., FERLIN, W. G. & LEPPER, H. (1995) Differential production of interferon-γ and interleukin-4 in response to Th1- and Th2-stimulating pathogens by $\gamma\delta$ T cells *in vivo*. *Nature*, **373**, 255–257.

FILLEY, W. V., HOLLEY, K. E., KEPHART, G. M. & GLEICH, G. J. (1982) Identification by immunofluorescence of eosinophil MBP in lung biopsies of patients with bronchial asthma. *Lancet*, **ii**, 11–16.

FINKELMAN, F. D., KATONA, I. M., URBAN, J. F., JR., SNAPPER, C. M., OHARA, J. & PAUL, W. E. (1986) Suppression of *in vivo* polyclonal IgE production by monoclonal antibody to the lymphokine B-cell stimulatory factor 1. *Proceedings of the National Academy of Sciences USA*, **83**, 9675–9678.

FINKELMAN, F. D., KATONA, I. M., MOSMANN, T. R. & COFFMAN, R. L. (1988a). IFN-γ regulates the isotypes of Ig secreted during *in vivo* humoral immune responses. *Journal of Immunology*, **140**, 1022–1027.

FINKELMAN, F. D., KATONA, I. M., URBAN, J. F., JR., HOLMES, J., OHARA, J., TUNG, A. S., SAMPLE, J. G. & PAUL, W. E. (1988b) IL-4 is required to generate and sustain *in vivo* IgE responses. *Journal of Immunology*, **141**, 2335–2341.

FOSTER, P. S., HOGAN, S. P., RAMSAY, A. J., MATTHAEI, K. I. & YOUNG, I. G. (1996) Interleukin 5 deficiency abolishes eosinophilia, airways hyperreactivity and lung damage in a mouse asthma model. *Journal of Experimental Medicine*, **183**, 195–201.

FOX, P. C. & SIRAGANIAN, R. P. (1981) IgE antibody suppression following aerosol exposure to antigens. *Immunology*, **43**, 227–234.

FREW, A. J. (1996) Cytokines, chemokines, T cells and allergy. *Clinical and Experimental Allergy*, **26**, 2–4.

FRIGAS, E., LOEGERING, D. A., SOLLEY, G. O., FARROW, G. M. & GLEICH, G. J. (1981) Elevated levels of eosinophil major basic protein in the sputum of patients with bronchial asthma. *Mayo Clinic Proceedings*, **56**, 345–353.

FUJIMAKI, H., NOHARA, O., ICHINOSE, T., WATANABE, N. & SAITO, S. (1994) IL-4 production in mediastinal lymph node cells in mice intratracheally instilled with diesel exhaust particulates and antigen. *Toxicology*, **92**, 261–268.

GAJEWSKI, T. F., PINNAS, M., WONG, T. & FITCH, F. W. (1991) Murine Th1 and Th2 clones proliferate optimally in response to distinct antigen-presenting cell populations. *Journal of Immunology*, **146**, 1750–1758.

GARSSEN, J., VANDEBRIEL, R. J., KIMBER, I. & VAN LOVEREN, H. (1996) Hypersensitivity reactions: definitions, basic mechanisms and localizations in VOS, J. G., YOUNES, M. & SMITH, E. (Eds) *Allergic Hypersensitivities Induced by Chemicals. Recommendations for Prevention*, pp. 19–58. Boca Raton: CRC Press.

GAVETT, S. H., O'HEARN, D. J., LI, X., HUANG, S-K., FINKELMAN, F. D. & WILLS-KARP, M. (1995) Interleukin 12 inhibits antigen-induced airway hyperresponsiveness, inflammation and Th2 cytokine expression in mice. *Journal of Experimental Medicine*, **182**, 1527–1536.

GERGEN, P. H., MULLALLY, D. I. & EVANS, R. (1988) National survey of prevalence of asthma among children in the United States. *Pediatrics*, **81**, 1–7.

GERRARD, J. W., HEINER, D. C., KO, C. G., MINK, J., MEYERS, A. & DOSMAN, J. A. (1980) Immunoglobulin levels in smokers and non-smokers. *Annals of Allergy*, **44**, 261–262.

GLEICH, G. J. (1990) The eosinophil and bronchial asthma: current understanding. *Journal of Allergy and Clinical Immunology*, **85**, 422–436.

GLEICH, G. J. & ADOLPHSON, C. R. (1986) The eosinophilic leukocyte: structure and function. *Advances in Immunology*, **39**, 177–253.
GOCINSKI, B. L. & TIGELAAR, R. E. (1990) Roles of CD4$^+$ and CD8$^+$ T cells in murine contact sensitivity revealed by *in vivo* monoclonal antibody depletion. *Journal of Immunology*, **144**, 4121–4128.
GRAMMER, L. C. (1993) Occupational immunologic lung disease, in PATTERSON, R., GRAMMER, L. C., GREENBERGER, P. A. & ZEISS, C. R. (Eds) *Allergic Diseases: Diagnosis and Management*, pp. 745–762, Philadelphia: Lippincott.
GRIFFITHS-JOHNSON, D. A., COLLINS, P. D., ROSSI, A. G., JOSE, P. J. & WILLIAMS, T. J. (1993) The chemokine, eotaxin, activates guinea pig eosinophils *in vitro* and causes their accumulation into the lung *in vivo*. *Biochemical and Biophysical Research Communications*, **197**, 1167–1172.
GULBENKIAN, A. R., EGAN, R. W., FERNANDEZ, X., JONES, H., KREUTNER, W., KUNG, T., PAGVARDI, F., SULLIVAN, L., ZURCHER, J. A. & WATNICK, A. S. (1992) Interleukin-5 modulates eosinophil accumulation in allergic guinea pig lung. *American Review of Respiratory Disease*, **146**, 263–265.
GUNDEL, R. H., LETTS, L. G. & GLEICH, G. J. (1991) Human eosinophil major basic protein induces airway constriction and airway hyperresponsiveness in primates. *Journal of Clinical Investigation*, **87**, 1470–1473.
HAHHTELA, T., LINHOLM, H., BJORKSTEN, F., KOSKENUVO, K. & LAITINEN, L. A. (1990) Prevalence of asthma in Finnish young men. *British Medical Journal*, **301**, 266–268.
HAMMARSTROM, L. & SMITH, C. I. E. (1987) Immunoglobulin subclass distribution of specific antibodies in allergic patients. *Allergy*, **42**, 529–536.
HOLLIDAY, M. R., COLEMAN, J. W., DEARMAN, R. J. & KIMBER, I. (1993) Induction of mast cell sensitization by chemical allergens: a comparative study. *Journal of Applied Toxicology*, **13**, 137–142.
HOLLIDAY, M. R., BANKS, E. M. S., DEARMAN, R. J., KIMBER, I. & COLEMAN, J. W. (1994) Interactions of IFN-γ with IL-3 and IL-4 in the regulation of serotonin and arachidonate release from mouse peritoneal mast cells. *Immunology*, **82**, 70–74.
HOLT, P. G. (1978) Inhibitory activity of unstimulated alveolar macrophages on T lymphocyte blastogenic response. *American Review of Respiratory Disease*, **118**, 791–793.
(1985) Downregulation of immune responses in the lower respiratory tract: role of alveolar macrophages. *Clinical and Experimental Immunology*, **63**, 261–270.
(1993) Macrophage: dendritic cell interaction in the regulation of the IgE response in asthma. *Clinical and Experimental Allergy*, **23**, 4–6.
HOLT, P. G., BATTY, J. E. & TURNER, J. K. (1981) Inhibition of specific IgE responses in mice by pre-exposure to inhaled antigen. *Immunology*, **42**, 409–417.
HOLT, P. G., DEGEBRODT, A., O'LEARY, C., KRSKA, K. & PLOZZA, T. (1985) T-cell activation by antigen-presenting cells from lung tissue digests: suppression by endogenous macrophages. *Clinical and Experimental Immunology*, **62**, 586–594.
HOLT, P. G., SCHON-HEGRAD, M. A. & OLIVER, J. (1988) MHC class II antigen-bearing dendritic cells in pulmonary tissues of the rat: regulation of antigen presentation activity by endogenous macrophage populations. *Journal of Experimental Medicine*, **167**, 262–274.
HOLT, P. G., SCHON-HEGRAD, M. A., PHILLIPS, M. J. & MCMENAMIN, P. G. (1989) Ia-positive dendritic cells form a tightly meshed network within the human

airway epithelium. *Clinical and Experimental Allergy*, **19**, 597–601.
HOLT, P. G., SCHON-HEGRAD, M. A. & MCMENAMIN, P. G. (1990a) Dendritic cells in the respiratory tract. *International Review of Immunology*, **6**, 139–149.
HOLT, P. G., SCHON-HEGRAD, M. A., OLIVER, J., HOLT, B. J. & MCMENAMIN, P. G. (1990b) A contiguous network of dendritic antigen-presenting cells within the respiratory epithelium. *International Archives of Allergy and Applied Immunology*, **91**, 155–159.
HSIEH, C-S., HEIMBERGER, A. B., GOLD, J. S., O'GARRA, A. & MURPHY, K. M. (1992) Differential regulation of T helper phenotype development by interleukins 4 and 10 in an $\alpha\beta$ T-cell-receptor transgenic system. *Proceedings of the National Academy of Sciences USA*, **89**, 6065–6069.
HSIEH, C-S., MACATONIA, S. E., TRIPP, C. S., WOLF, S. F., O'GARRA, A. & MURPHY, K. M. (1993) Development of Th1 CD4$^+$ T cells through IL-12 produced by *Listeria*-induced macrophages. *Science*, **260**, 547–549.
HUMBERT, M. (1996) Pre-eosinophilic cytokines in asthma. *Clinical and Experimental Allergy*, **26**, 123–127.
HUSSAIN, R. & OTTESEN, E. A. (1986) IgE responses in human filariasis. IV. Parallel antigen recognition by IgE and IgG4 subclass antibodies. *Journal of Immunology*, **136**, 1859–1863.
INABA, K. & STEINMAN, R. M. (1984) Resting and sensitized T lymphocytes exhibit distinct (antigen presenting cell) requirements for growth and lymphokine release. *Journal of Experimental Medicine*, **160**, 1717–1735.
ISHIZAKA, A., SAKIYAMA, Y., NAKANISHI, M., TOMIZAWA, K., OSHIKA, E., KOJIMA, K., TAGUCHI, Y., KANDIL, E. & MATSUMOTO, S. (1990) The inductive effect of interleukin-4 on IgG4 and IgE synthesis in human peripheral blood lymphocytes. *Clinical and Experimental Immunology*, **79**, 392–396.
IWAMI, T., NAGAI, H., TSURUOKA, N. & KODA, A. (1993) Effect of murine recombinant interleukin-5 on bronchial reactivity in guinea-pigs. *Clinical and Experimental Allergy*, **23**, 32–38.
IWAMOTO, I. & TAKATSU, K. (1995) Evaluation of airway hyperreactivity in interleukin-5 transgenic mice. *International Archives of Allergy and Immunology*, **108**, 28–30.
IWAMOTO, I., NAKAJIMA, H., ENDO, H. & YOSHIDA, S. (1993) Interferon γ regulates antigen-induced eosinophil recruitment into the mouse airways by inhibiting the infiltration of CD4$^+$ T cells. *Journal of Experimental Medicine*, **177**, 573–576.
JOSE, P. H., GRIFFITHS-JOHNSON, D. A., COLLINS, P. D., WALSH, D. T., MOQBEL, R., TOTTY, N. F., TRUONG, O., HSUAN, J. J. & WILLIAMS, T. J. (1994) Eotaxin: a potent eosinophil chemoattractant cytokine detected in a guinea pig model of allergic airways inflammation. *Journal of Experimental Medicine*, **179**, 881–887.
KAMOGAWA, Y., MINASI, L. E., CARDING, S. R., BOTTOMLY, K. & FLAVELL, R. (1993) The relationship of IL-4- and IFN-γ-producing T cells by lineage ablation of IL-4-producing cells. *Cell*, **75**, 985–995.
KAPSENBERG, M. L., WIERENGA, E. A., BOS, J. D. & JANSEN, H. M. (1991) Functional subsets of allergen-reactive CD4$^+$ T cells. *Immunology Today*, **12**, 392–395.
KAPSENBERG, M. L., WIERENGA, E. A., STIEKMA, F. E. M., TIGGELMAN, A. M. B. C. & BOS, J. D. (1992) T_H1 lymphokine production profiles of nickel-specific CD4$^+$ T lymphocyte clones from nickel contact allergic and non-allergic

individuals. *Journal of Investigative Dermatology*, **98**, 59–63.
KAROL, M. H. (1992) Occupational asthma and allergic reactions to inhaled compounds, in MILLER, K., TURK, J. & NICKLIN, S. (Eds) *Principles and Practice of Immunotoxicology*, pp. 228–241, Oxford: Blackwell Scientific Publications.
KAROL, M. H., TOLLERUD, D. J., CAMPBELL, T. P., FABBRI, L., MAESTRELLI, P., SAETTA, M. & MAPP, C. E. (1994) Predictive value of airways hyperresponsiveness and circulating IgE for identifying types of responses to toluene diisocyanate inhalation challenge. *American Journal of Respiratory and Critical Care Medicine*, **149**, 611–615.
KELSO, A. (1995) Th1 and Th2 subsets: paradigms lost? *Immunology Today*, **16**, 374–379.
KEMENY, D. M. & DIAZ-SANCHEZ, D. (1993) The role of $CD8^+$ T cells in the regulation of IgE. *Clinical and Experimental Allergy*, **23**, 466–470.
KEMENY, D. M., NOBLE, A., DIAZ-SANCHEZ, D., STAYNOV, D. & LEE, T. (1992) Ricin-sensitive early activated $CD8^+$ T cells suppress IgE responses and regulate the production of IFN-γ and IL-4 by splenic $CD4^+$, $CD8^+$ T cells. *International Archives of Allergy and Immunology*, **99**, 362–365.
KEMENY, D. M., NOBLE, A., HOLMES, B. J. & DIAZ-SANCHEZ, D. (1994) Immune regulation: a new role for the $CD8^+$ T cell. *Immunology Today*, **15**, 107–110.
KIMATA, H., YOSHIDA, A., ISHIOKA, C., LINDLEY, I. & MIKAWA, H. (1992) Interleukin 8 (IL-8) actively inhibits immunoglobulin E production induced by IL-4 in human B cells. *Journal of Experimental Medicine*, **176**, 1227–1231.
KIMBER, I. (1995) Mechanisms of pulmonary sensitization, in THOMAS, H., HESS, R. & WAECHTER, F. (Eds). *Toxicology of Industrial Compounds*, pp. 141–151. London: Taylor & Francis.
(1996a) Mechanisms of chemical respiratory allergy, in VOS, J. G., YOUNES, M. & SMITH, E. (Eds) *Allergic Hypersensitivities Induced by Chemicals. Recommendations for Prevention*, pp. 59–75. Boca Raton: CRC Press.
(1996b) Chemical respiratory allergy: exposure and dose–response relationships, in VOS, J. G., YOUNES, M. & SMITH, E. (Eds) *Allergic Hypersensitivities Induced by Chemicals. Recommendations for Prevention*, pp. 139–147. Boca Raton: CRC Press.
(1996c) The role of the skin in the development of chemical respiratory hypersensitivity. *Toxicology Letters* (in press).
KIMBER, I. & CUMBERBATCH, M. (1992) Dendritic cells and cutaneous immune responses to chemical allergens. *Toxicology and Applied Pharmacology*, **117**, 137–146.
KIMBER, I. & DEARMAN, R. J. (1992) The mechanisms and evaluation of chemically induced allergy. *Toxicology Letters*, **64/65**, 79–84.
(1994) Immune responses to contact and respiratory allergens, in DEAN, J. H., LUSTER, M. I., MUNSON, A. E. & KIMBER, I. (Eds) *Immunotoxicology and Immunopharmacology*, 2nd Edn, pp. 663–679, New York: Raven Press.
KOHLER, J., MARTIN, S., PFLUGFELDER, U., RUH, H., VOLLMER, J. & WELTZIEN, H. U. (1995) Cross-reactive trinitrophenylated peptides as antigens for class II major histocompatibility complex-restricted T cells and inducers of contact sensitivity in mice. *European Journal of Immunology*, **25**, 92–101.
KOLBE, L., HEUSSER, C. & KOLSCH, E. (1994) Epitope-dependent nonreciprocal regulation of IgE and IgG2a antibody formation. *International Archives of Allergy*

and Immunology, 103, 214–216.
Kopf, M., Le Gros, G., Bachmann, M., Lamers, M. C., Bleuthmann, H. & Kohler, G. (1993) Disruption of the murine IL-4 gene blocks Th2 cytokine responses. Nature, 362, 245–248.
Kuchroo, V. K., Das, M. P., Brown, J. A., Ranger, A. M., Zamvil, S. S., Sobel, R. A., Weiner, H. L., Nabavi, N. & Glimcher, L. H. (1995) B7-1 and B7-2 costimulatory molecules activate differentially the Th1/Th2 developmental pathways: application to autoimmune disease therapy. Cell, 80, 707–718.
Kuhn, R., Rajewsky, K. & Muller, W. (1991) Generation and analysis of interleukin-4 deficient mice. Science, 254, 707–710.
Kumar, V., Bhardwaj, V., Soares, L., Alexander, J., Sette, A. & Sercarz, E. (1995) Major histocompatibility complex binding affinity of an antigenic determinant is crucial for the differential secretion of interleukin 4/5 or interferon γ by T cells. Proceedings of the National Academy of Sciences USA, 92, 9510–9514.
Kung, T. T., Stelts, D. M., Zurcher, J. A., Jones, H., Umland, S. P., Egan, R. W., Kreutner, W. & Chapman, R. W. (1995) Interferon-γ and antibodies to interleukin-5 and interleukin-4 inhibit the pulmonary eosinophilia in allergic mice. Inflammation Research, 2, S185–S186.
Le Gros, G. & Erard, F. (1994) Non-cytotoxic, IL-4, IL-5, IL-10 producing $CD8^+$ T cells: their activation and effector functions. Current Opinion in Immunology, 6, 453–457.
Lipscomb, M. F., Bice, D. E., Lyons, C. R., Schuyler, M. R. & Wilkes, D. (1995) The regulation of pulmonary immunity. Advances in Immunology, 59, 369–455.
Lopez, A. F., Sanderson, C. J., Gamble, J. R., Campbell, H. D., Young, I. G. & Vadas, M. A. (1988) Recombinant interleukin 5 is a selective activator of human eosinophil function. Journal of Experimental Medicine, 167, 219–224.
Lukacs, N. W., Strieter, R. M., Chensue, S. W. & Kunkel, S. L. (1994) Interleukin-4-dependent pulmonary eosinophil infiltration in a murine model of asthma. American Journal of Respiratory Cell and Molecular Biology, 10, 526–532.
Macatonia, S. E., Doherty, T. M., Knight, S. C. & O'Garra, A. (1993) Differential effect of IL-10 on dendritic cell-induced T cell proliferation and IFN-γ production. Journal of Immunology, 150, 3755–3765.
Magnusson, C. G. (1986) Maternal smoking influences cord serum IgE levels and increases the risk of subsequent infant allergy. Journal of Allergy and Clinical Immunology, 78, 898–904.
Martin, S. & Weltzien, H. U. (1994) T cell recognition of haptens: a molecular view. International Archives of Allergy and Immunology, 104, 10–16.
Martin, S., von Bonin, A., Fessler, C., Pflugfelder, U. & Weltzien, H. U. (1993) Structural complexity of antigenic determinants for class I MHC-restricted, hapten-specific T cells: two qualitatively differing types of $H-2K^b$-restricted TNP epitopes. Journal of Immunology, 151, 678–687.
McKnight, A. J., Zimmer, G. J., Fogelman, I., Wolf, S. F. & Abbas, A. K. (1994) Effects of IL-12 on helper T cell-dependent immune responses in vivo. Journal of Immunology, 152, 2172–2179.
McMenamin, C. & Holt, P. G. (1993) The natural immune response to inhaled soluble protein antigens involves major histocompatibility complex (MHC) class I-restricted $CD8^+$ T cell-mediated but MHC class II-restricted $CD4^+$ T cell-

dependent immune deviation resulting in selective suppression of immunoglobulin E production. *Journal of Experimental Medicine*, **178**, 889–899.

MCMENAMIN, C., PIMM, C., MCKERSEY, M. & HOLT, P. G. (1994) Regulation of IgE responses to inhaled antigen in mice by antigen-specific $\gamma\delta$ T cells. *Science*, **265**, 1869–1871.

MCMENAMIN, C., MCKERSEY, M., KUHNLEIN, P., HUNIG, T. & HOLT, P. G. (1995) $\gamma\delta$ T cells down-regulate primary IgE responses in rats to inhaled soluble protein antigens. *Journal of Immunology*, **154**, 4390–4394.

MCWILLIAM, A. S., NELSON, D., THOMAS, J. A. & HOLT, P. G. (1994) Rapid dendritic cell recruitment is a hallmark of the acute inflammatory response at mucosal surfaces. *Journal of Experimental Medicine*, **179**, 1331–1336.

MINTY, A., CHALON, P., DEROCQ, J.-M., DUMONT, X., GUILLEMOT, J-C., KAGHAD, M., LABIT, C., LEPLATOIS, P., LIAUZUN, P. & MILOUX, B. (1993) Interleukin-13 is a new human lymphokine regulating inflammatory and immune responses. *Nature*, **362**, 248–250.

MOLLER, D. R., GALLAGHER, J. S., BERNSTEIN, D. I., WILCOX, T. G., BURROUGHS, H. E. & BERNSTEIN, I. L. (1985) Detection of IgE-mediated respiratory hypersensitivity in workers exposed to hexahydrophthalic anhydride. *Journal of Allergy and Clinical Immunology*, **76**, 663–672.

MORRIS, S. C., MADDEN, K. B., ADAMOVICZ, J. L., GAUSE, W. C., HUBBARD, B. R., GATELY, M. K. & FINKELMAN, F. D. (1994) Effects of IL-12 on in vivo cytokine gene expression and Ig isotype selection. *Journal of Immunology*, **152**, 1047–1056.

MOSMANN, T. R. & COFFMAN, R. L. (1989) Heterogeneity of cytokine secretion patterns and functions of helper T cells. *Advances in Immunology*, **46**, 111–145.

MOSMANN, T. R. & SAD, S. (1996) The expanding universe of T-cell subsets: Th1, Th2 and more. *Immunology Today*, **17**, 138–146.

MOSMANN, T. R., CHERWINSKI, H., BOND, M. W., GIEDLIN, M. A. & COFFMAN, R. L. (1986) Two types of murine helper T cell clone. 1. Definition according to profiles of lymphokine activities and secreted proteins. *Journal of Immunology*, **136**, 2348–2357.

MOSMANN, T. R., SCHUMACHER, J. H., STREET, N. F., BUDD, R., O'GARRA, A., FONG, T. A. T., BOND, M. W., MOORE, K. W. M., SHER, A. & FIORENTINO, D. F. (1991) Diversity of cytokine synthesis and function of mouse CD4$^+$ T cells. *Immunological Reviews*, **123**, 209–229.

MURANKA, M., SUZUKI, S., KOIZUMI, K., TAKAFUJI, S., MIYAMOTO, T., IKEMORI, R. & TOKIWA, H. (1986) Adjuvant activity of diesel-exhaust particulates for the production of IgE antibody in mice. *Journal of Allergy and Clinical Immunology*, **77**, 616–623.

NEWMAN TAYLOR, A. J. (1988) Occupational asthma. *Postgraduate Medical Journal*, **64**, 505–510.

(1995) Environmental determinants of asthma. *Lancet*, **345**, 296–299.

NIELSEN, J., WELINDER, H., OTTOSON, H., BENSRYD, I., VENGE, P. & SKERFVING, S. (1994) Nasal challenge shows pathogenetic relevance of specific IgE serum antibodies for nasal symptoms caused by hexahydrophthalic anhydride. *Clinical and Experimental Allergy*, **24**, 440–449.

NOBLE, A. & KEMENY, D. M. (1995) Interleukin-4 and interferon-γ regulate differentiation of CD8$^+$ T cells into populations with divergent cytokine profiles. *International Archives of Allergy and Immunology*, **107**, 186–188.

NOBLE, A., MACARY, P. A. & KEMENY, D. M. (1995) IFN-γ and IL-4 regulate the growth and differentiation of CD8$^+$ T cells into subpopulations with distinct cytokine profiles. *Journal of Immunology*, **155**, 2928–2937.

OMENAAS, E., BAKKE, P., ELSAYED, S., HANOA, R. & GULSVIK, A. (1994) Total and specific serum IgE levels in adults: relationship to sex, age and environmental factors. *Clinical and Experimental Allergy*, **24**, 530–539.

PARK, H. S. & HONG, C-S. (1991) The significance of specific IgG and IgG4 antibodies to a reactive dye in exposed workers. *Clinical and Experimental Allergy*, **21**, 357–362.

PARRONCHI, P., MACCHIA, D., PICCINNI, M-P., BISWAS, P., SIMONELLI, C., MAGGI, E., RICCI, M., ANSARI, A. A. & ROMAGNANI, S. (1991) Allergen and bacterial antigen-specific T-cell clones established from atopic donors show a different profile of cytokine production. *Proceedings of the National Academy of Sciences USA*, **88**, 4538–4542.

PENE, J., ROUSSET, F., BRIERE, F., CHRETIEN, I., PALIARD, X., BANCHEREAU, J., SPITS, H. & DE VRIES, J. (1988) IgE production by normal human B cells induced by alloreactive T cell clones is mediated by IL-4 and suppressed by IFN-γ. *Journal of Immunology*, **141**, 1218–1224.

PFEIFFER, C., STEIN, J., SOUTHWOOD, S., KETELAAR, H., SETTE, A. & BOTTOMLY, K. (1995) Altered peptide ligands can control CD4 T lymphocyte differentiation *in vivo*. *Journal of Experimental Medicine*, **181**, 1569–1574.

POLLARD, A. M. & LIPSCOMB, M. F. (1990) Characterization of murine lung dendritic cells: similarities to Langerhans cells and thymic dendritic cells. *Journal of Experimental Medicine*, **172**, 159–167.

PUNNONEN, J., AVERSA, G., COCKS, B. G., MCKENZIE, A. N. J., MENON, S., ZURAWSKI, G., DE WAAL MALEFYT, R. & DE VRIES, J. E. (1993a) Interleukin 13 induces interleukin 4-independent IgG4 and IgE synthesis and CD23 expression by human B cells. *Proceedings of the National Academy of Sciences USA*, **90**, 3730–3734.

PUNNONEN, J., PUNNONEN, K., JANSEN, C. T. & KALIMO, K. (1993b) Interferon (IFN)-α, IFN-γ, interleukin (IL)-2 and arachidonic acid metabolites modulate IL-4-induced IgE synthesis similarly in healthy persons and in atopic dermatitis patients. *Allergy*, **48**, 189–195.

REINER, S. L. & SEDER, R. A. (1995) T helper cell differentiation in immune response. *Current Opinion in Immunology*, **7**, 360–366.

RENZ, H., LACK, G., SALOGA, J., SCHWINZER, R., BRADLEY, K., LOADER, J., KUPFER, A., LARSEN, G. L. & GELFAND, E. W. (1994) Inhibition of IgE production and normalization of airways responsiveness by sensitized CD8 T cells in mouse model of allergen-induced sensitization. *Journal of Immunology*, **152**, 351–360.

RENZ, H., BRADLEY, K., ENSSLE, K., LOADER, J. E., LARSEN, G. L. & GELFAND, E. W. (1996) Prevention of the development of immediate hypersensitivity and airway hyperresponsiveness following *in vivo* treatment with soluble IL-4 receptor. *International Archives of Allergy and Immunology*, **109**, 167–176.

RESNICK, M. B. & WELLER, P. F. (1993) Mechanisms of eosinophil recruitment. *American Journal of Respiratory Cell and Molecular Biology*, **8**, 349–355.

RICCI, M., ROSSI, O., BERTONI, M. & MATUCCI, A. (1993) The importance of Th2-like cells in the pathogenesis of airway allergic inflammation. *Clinical and Experimental Allergy*, **23**, 360–369.

RIEMANN, H., SCHWARZ, A., GRABBE, S., ARAGNE, Y., LUGER, T. A., WYSOCKA, M., KUBIN, M., TRINCHIERI, G. & SCHWARZ, T. (1996) Neutralization of IL-12 *in vivo* prevents induction of contact hypersensitivity and induces hapten-specific tolerance. *Journal of Immunology*, 156, 1799–1803.

ROBINSON, D. S. (1993) Interleukin-5, eosinophils and bronchial hyperreactivity. *Clinical and Experimental Allergy*, 23, 1–3.

ROBINSON, D. S., HAMID, Q., YING, S., TSICOPOULOS, A., BARKANS, J., BENTLEY, A. M., CORRIGAN, C., DURHAM, S. R. & KAY, A. B. (1992) Predominant Th2-like bronchoalveolar T-lymphocyte population in atopic asthma. *New England Journal of Medicine*, 326, 298–304.

ROBINSON, D. S., HAMID, Q., BENTLEY, A., YING, S., KAY, A. B. & DURHAM, S. R. (1993) Activation of $CD4^+$ T cells, increased Th2-type cytokine mRNA expression and eosinophil recruitment in bronchoalveolar lavage after allergen inhalation challenge in patients with atopic asthma. *Journal of Allergy and Clinical Immunology*, 92, 313–324.

ROCHE, W. R., BEASLEY, R., WILLIAMS, J. H. & HOLGATE, S. T. (1989) Subepithelial fibrosis in the bronchi of asthmatics. *Lancet*, i, 520–524.

ROMAGNANI, S. (1991) Human T_{H1} and T_{H2} subsets: doubt no more. *Immunology Today*, 12, 256–257.

(1992) Induction of T_{H1} and T_{H2} responses: a key role for the 'natural' immune response. *Immunology Today*, 13, 379–381.

ROMAGNANI, S., DEL PRETE, G., MAGGI, E., PARRONCHI, P., TIRI, A., MACCHIA, D., GIUDIZI, M. G., ALMERIGOGNA, F., RICCI, M. & BIAGIOTTI, R. (1989) Role of interleukins in induction and regulation of human IgE synthesis. *Clinical Immunology and Immunopathology*, 50, S13–S23.

ROMAGNANI, S., DEL PRETE, G., MAGGI, E., PARRONCHI, P., DE CARLI, M., MACCHIA, D., MANETTI, R., SAMPOGNARO, S., PICCINNI, M.-P. & GIUDIZI, M. G. (1992) Human T_{H1} and T_{H2} subsets. *International Archives of Allergy and Immunology*, 99, 242–245.

ROTHENBERG, M. E., LUSTER, A. D., LILLY, C. M., DRAZEN, J. M. & LEDER, P. (1995) Constitutive and allergen-induced expression of eotaxin mRNA in the guinea pig lung. *Journal of Experimental Medicine*, 181, 1211–1216.

SAD, S., MARCOTTE, R. & MOSMANN, T. R. (1995) Cytokine-induced differentiation of precursor mouse $CD8^+$ T cells into cytotoxic $CD8^+$ T cells secreting Th1 or Th2 cytokines. *Immunity*, 2, 271–279.

SAITO, S., DORF, M. E., WATANABE, M. & TADAKUMA, T. (1994) Preferential induction of IL-4 is determined by the type and duration of antigenic stimulation. *Cellular Immunology*, 153, 1–8.

SALVAGGIO, J. E., BUTCHER, B. T. & O'NEIL, C. E. (1988) Occupational asthma due to chemical agents. *Journal of Allergy and Clinical Immunology*, 78, 1053–1057.

SCHLEIMER, P., STERBINSKY, S. A., KAISER, J., BICKEL, C. A., KLUNK, D. A., TOMIOKA, A., NEWMAN, W., LUSCINSKAS, F. W., GIMBRONE, M. A. JR & MCINTYRE, B. W. (1992) IL-4 induces adherence of human eosinophils and basophils, but not neutrophils, to endothelium: association with expression of VCAM-1. *Journal of Immunology*, 148, 1086–1092.

SCHMITT, E., HOEHN, P., HUELS, C., GOEDERT, S., PALM, N., RUDE, E. & GERMAN, T. (1994) T helper type 1 development of naive $CD4^+$ T cells requires the co-ordinate action of interleukin-12 and interferon-γ and is inhibited by transforming growth factor-β. *European Journal of Immunology*, 24, 793–798.

SCHON-HEGRAD, M. A., OLIVER, J., MCMENAMIN, P. G. & HOLT, P. G. (1991) Studies on the density, distribution and surface phenotype of intraepithelial class II MHC antigen (Ia)-bearing dendritic cells (DC) in the conducting airways. *Journal of Experimental Medicine*, **173**, 1345–1356.

SEDER, A., GAZZINELLI, R., SHER, A. & PAUL, W. E. (1993) Interleukin 12 acts directly on CD4$^+$ T cells to enhance priming for interferon-γ production and diminishes interleukin-4 inhibition of such priming. *Proceedings of the National Academy of Sciences USA*, **90**, 10188–10192.

SEDER, R. A. & LE GROS, G. G. (1995) The functional role of CD8$^+$ T helper type 2 cells. *Journal of Experimental Medicine*, **181**, 5–7.

SEDGWICK, J. & HOLT, P. G. (1985) Down-regulation of immune responses to inhaled antigen. Studies on the mechanism of induced suppression. *Immunology*, **56**, 635–642.

SEMINARIO, M.-C. & GLEICH, G. J. (1994) The role of eosinophils in the pathogenesis of asthma. *Current Opinion in Immunology*, **6**, 860–864.

SERTL, K., TAKEMURA, T., TSCHACHLER, E., FERRANS, V. J., KALINER, M. A. & SHEVACH, E. M. (1986) Dendritic cells with antigen-presenting capability reside in airway epithelium, lung parenchyma and visceral pleura. *Journal of Experimental Medicine*, **163**, 436–451.

SMITH, C. A. & RENNICK, D. M. (1986) Characterization of a murine lymphokine distinct from interleukin 3 (IL-3) possessing a T-cell growth factor activity and a mast cell growth factor activity that synergizes with IL-3. *Proceedings of the National Academy of Sciences USA*, **83**, 1857–1861.

SWAIN, S. L., BRADLEY, L. M., CROFT, M., TONKONOGY, S., ATKINS, G, WEINBERG, A. D., DUNCAN, D. D., HEDRICK, S. M., DUTTEN, R. W. & HUSTON, G. (1991) Helper T-cell subsets: phenotype, function and the role of lymphokines in regulating their development. *Immunological Reviews*, **123**, 115–144.

TAKAFUJI, S., SUZUKI, S., KOIZUMI, K., TADOKORO, K., MIYAMOTO, T., IKEMORI, R. & MURANKA, M. (1987) Diesel-exhaust particulates inoculated by the intranasal route have an adjuvant activity for IgE production in mice. *Journal of Allergy and Clinical Immunology*, **79**, 639–645.

TAKENAKA, H., ZHANG, K., DIAZ-SANCHEZ, D., TSIEN, A. & SAXON, A. (1995) Enhanced human IgE production results from exposure to the aromatic hydrocarbons from diesel exhaust: direct effects on B-cell IgE production. *Journal of Allergy and Clinical Immunology*, **95**, 103–115.

TARKOWSKI, M. & GORSKI, P. (1995) Increased IgE antiovalbumin level in mice exposed to formaldehyde. *International Archives of Allergy and Immunology*, **106**, 422–424.

TEPPER, R. I., LEVINSON, D. A., STANGER, B. Z., CAMPOS-TORRES, J., ABBAS, A. K. & LEDER, P. (1990) IL-4 induces allergic-like inflammatory disease and alters T cell development in transgenic mice. *Cell*, **62**, 457–467.

THEPEN, T., VAN ROOIJEN, N. & KRAAL, G. (1989) Alveolar macrophage elimination *in vivo* is associated with an increase in pulmonary immune response in mice. *Journal of Experimental Medicine*, **170**, 499–509.

THEPEN, T., MCMENAMIN, C., OLIVER, J., KRAAL, G. & HOLT, P. G. (1991) Regulation of immune responses to inhaled antigen by alveolar macrophages: differential effects of *in vivo* alveolar macrophage elimination on the induction of tolerance versus immunity. *European Journal of Immunology*, **21**, 2845–2850.

Thepen, T., McMenamin, C., Girn, B., Kraal, G. & Holt, P. G. (1992) Regulation of IgE production in presensitized animals: in vivo elimination of alveolar macrophages selectively increases IgE responses to inhaled allergen. *Clinical and Experimental Allergy*, **22**, 1107–1114.

Thompson-Snipes, L., Dhar, V., Bond, M. W., Mosmann, T. R., Moore, K. W. & Rennick, D. M. (1991) Interleukin-10: a novel stimulatory factor for mast cells and their progenitors. *Journal of Experimental Medicine*, **173**, 507–510.

Topping, M. D., Forster, H. W., Ide, C. W., Kennedy, F. M., Leach, A. M. & Sorkin, S. (1989) Respiratory allergy and specific immunoglobulin E and immunoglobulin G antibodies to reactive dyes used in the wool industry. *Journal of Occupational Medicine*, **31**, 859–862.

Trinchieri, G. & Scott, P. (1994) The role of interleukin 12 in the immune response, disease and therapy. *Immunology Today*, **15**, 460–463.

Underwood, S. L., Kemeny, D. M., Lee, T. H., Raeburn, D. & Karlsson, J-A. (1995) IgE production, antigen-induced airway inflammation and airway hyperreactivity in the Brown Norway rat: the effects of ricin. *Immunology*, **85**, 256–261.

van der Heijden, F. L., Wierenga, E. A., Bos, J. D. & Kapsenberg, M. L. (1991) High frequency of IL-4-producing CD4$^+$ allergen-specific T lymphocytes in atopic dermatitis lesional skin. *Journal of Investigative Dermatology*, **97**, 389–394.

Venables, K. M., Topping, M. D., Howe, W., Luczynska, C. M., Hawkins, R. & Newman Taylor, A. J. (1985) Interaction of smoking and atopy in producing specific IgE antibody against a hapten–protein conjugate. *British Medical Journal*, **290**, 201–204.

Wang, J. M., Rambaldi, A., Biondi, A., Chen, Z. G., Sanderson, C. J. & Mantovani, A. (1989) Recombinant interleukin 5 is a selective eosinophil chemoattractant. *European Journal of Immunology*, **19**, 701–705.

Wardlaw, A. J. (1993) The role of air pollution in asthma. *Clinical and Experimental Allergy*, **23**, 81–96.

Weaver, C. T., Hawrylowicz, C. M. & Unanue, E. R. (1988) T helper cell subsets require the expression of distinct costimulatory signals by antigen-presenting cells. *Proceedings of the National Academy of Sciences USA*, **85**, 8181–8185.

Weller, P. (1991) The immunobiology of eosinophils. *New England Journal of Medicine*, **324**, 1110–1118.

Weitzman, M., Gortmaker, S. I., Sobol, A. M. & Perrin, J. M. (1992) Recent trends in the prevalence and severity of childhood asthma. *Journal of the American Medical Association*, **268**, 2673–2677.

Weltzien, H. U., Moulon, C., Martin, S., Padovan, E., Hartmann, U. & Kohler, J. (1996) T cell immune responses to haptens. Structural models for allergic and autoimmune reactions. *Toxicology*, **107**, 141–151.

Wenner, C. A., Guler, M. L., Macatonia, S. E., O'Garra, A. & Murphy, K. M. (1996) Roles of IFN-γ and IFN-α in IL-12-induced T helper cell-1 development. *Journal of Immunology*, **156**, 1442–1447.

Xia, W., Pinto, C. E. & Kradin, R. L. (1995) The antigen-presenting activities of Ia$^+$ dendritic cells shift dynamically from lung to lymph node after an airway challenge with soluble antigens. *Journal of Experimental Medicine*, **181**, 1275–1283.

Xu, H., Dilulio, N. A. & Fairchild, R. L. (1996) T cell populations primed by hapten sensitization in contact sensitivity are distinguished by polarized patterns of cytokine production. Interferon-γ-producing (Tc1) effector CD8$^+$ T cells and interleukin (IL) 4/IL-10-producing (Th2) negative regulatory CD4$^+$ T cells. *Journal of Experimental Medicine*, **183**, 1001–1012.

YAMAGUCHI, Y., SUDA, T., SUDA, J., EGUCHI, M., MIURA, Y., HARAD, M., TAMINAGA, A. & TAKATSU, K. (1988) Purified interleukin 5 supports the terminal differentiation and proliferation of murine eosinophilic precursors. *Journal of Experimental Medicine*, **167**, 43–56.

YING, S., DURHAM, S. R., JACOBSON, M. R., RAK, S., MASUYAMA, K., LOWHAGAN, O., KAY, A. B. & HAMID, Q. A. (1994) T lymphocytes and mast cells express messenger mRNA for interleukin-4 in the nasal mucosa in allergen-induced rhinitis. *Immunology*, **82**, 200–206.

YOKOTA, T., COFFMAN, R. L., HAGIWARA, H., RENNICK, D. M., TAKEBE, Y., YOKOTA, K., GEMMELL, L., SCHRADER, B., YANG, G. & MEYERSON, P. (1987) Isolation and characterization of lymphokine cDNA clones encoding mouse and human IgA-enhancing and eosinophil-colony stimulating factor activities. Relationship to interleukin 5. *Proceedings of the National Academy of Sciences USA*, **84**, 7388–7392.

YOSHIMOTO, T., BENDELAC, A., HU-LI, J. & PAUL, W. E. (1995a) Defective IgE production by SJL mice is linked to the absence of CD4$^+$ NK1.1$^+$ T cells that promptly produce interleukin 4. *Proceedings of the National Academy of Sciences USA*, **92**, 11931–11934.

YOSHIMOTO, T., BENDELAC, A., WATSON, C., HU-LI, J. & PAUL, W. E. (1995b) Role of NK1.1$^+$ T cells in a Th2 response and in immunoglobulin E production. *Science*, **270**, 1845–1847.

ZETTERSTROM, O., OSTERMAN, K., MACHADO, L. & JOHANSSON, S. G. (1981) Another smoking hazard: raised serum IgE concentration and increased risk of occupational allergy. *British Medical Journal*, **283**, 1215–1217.

6

Predictive Assessment of Respiratory Sensitizing Potential in Guinea Pigs

KATHERINE SARLO and HARRY L. RITZ

The Procter & Gamble Company, Cincinnati, Ohio

6.1 Introduction

Respiratory allergic responses are adverse physiological events that develop upon exposure to a variety of agents. These agents may be high-molecular-weight compounds such as proteins or low-molecular-weight (LMW) compounds such as reactive chemicals. The large majority of respiratory allergic reactions in humans are caused by IgE-mediated antibody responses, although other types of immune mechanisms may play a role in responses to certain LMW chemicals. Allergic responses may occur as upper airway symptoms (i.e. rhinitis, conjunctivitis) or lower airway symptoms (i.e. asthma) and, in some cases, may be life-threatening. The prevalence of occupational asthma has been reported to range from 2 to 15 per cent of the workforce; these rates include both immune and non-immune causes of asthma (Chan-Yeung, 1990).

Respiratory allergic responses can be classified as immediate onset or late onset, depending upon the time of occurrence after allergen exposure. Immediate-onset reactions usually occur within a few minutes after exposure, whereas late-onset reactions occur hours after exposure. Some individuals will exhibit both types of response; in these cases, they are described as dual responses. Immediate and/or late-onset responses can occur after exposure to proteins or reactive chemicals, although late-onset responses are most frequently observed after exposure to LMW chemicals.

Several animal models have been developed to study mechanisms of respiratory allergic responses to proteins and chemicals. However, only a few of these have been used as predictive tests in hazard identification and risk assessment. These models have employed guinea pigs, mice or rats. This report describes guinea pig models developed as predictive tests for risk assessment of proteins and LMW chemicals as respiratory allergens.

6.2 The Guinea Pig as an Animal Model for Respiratory Allergy

The guinea pig has been used as an animal model for respiratory allergy and anaphylaxis for many decades (Ratner et al., 1927; Paupe, 1984). As with humans, the lung is a major shock organ for anaphylactic responses to allergens. Guinea pigs exhibit pulmonary responses upon exposure to histamine and experience both immediate-onset and late-onset reactions upon exposure to allergens (Karol et al., 1981b; Griffith-Johnson et al., 1993). Airway hyperreactivity after pulmonary reaction to allergen has also been demonstrated in the guinea pig (Griffith-Johnson and Karol, 1991). Eosinophilic influx, one hallmark of asthma in man, has also been shown to occur in the guinea pig (Lapa e Silva et al., 1993, 1994). Agents that block eosinophilic influx also block allergen-stimulated airway hyperreactivity in the guinea pig (Pretolani et al., 1994; van Oosterhout et al., 1995).

The major allergic antibody in the guinea pig is IgG1, whereas the major allergic antibody in man is IgE. Eosinophilic inflammatory responses noted after allergen challenge of guinea pigs have been associated with allergen-specific IgG1 antibody. This has been demonstrated in the respiratory tract (Griffith-Johnson et al., 1993) as well as in the peritoneal cavity and skin (Parish, 1972). An *in vivo* passive cutaneous anaphylaxis (PCA) test has been used for many years to measure guinea pig IgG1 antibody. The need to conduct PCA tests has added to the expense of guinea pig experimentation, as well as increasing the numbers of animals required for assessment of allergenicity. However, with the recent availability of reagents that can specifically recognize guinea pig IgG1 antibody, the PCA test has been replaced with the *in vitro* enzyme-linked immunosorbent assay (ELISA) (Prento, 1991; Kawabata et al., 1995).

Guinea pigs may also produce IgE antibody in small amounts after allergen exposure (Dobson et al., 1971), but the level of allergen-specific IgE is generally lower than that of IgG1. Guinea pig IgE can be detected in a 4- to 7-day PCA test and will cause the same type of hypersensitivity responses as IgG1 antibody. The presence of IgE antibodies to ovalbumin has been associated with late-onset pulmonary reactions in the guinea pig. The relationship between IgG1 antibody and IgE antibody and the development of pulmonary reactivity is not fully understood in this species. Due to a sparsity of reagents to identify immune and inflammatory cells in the guinea pig, it has been difficult to conduct in-depth mechanistic experimentation to understand antibody and pulmonary responses in this species. This disadvantage of the guinea pig as a test subject will become less important as these reagents become available.

Guinea pigs respond to a wide range of protein and chemical allergens and dose-dependent allergic antibody and pulmonary responses can be shown to occur upon exposure to a variety of allergens. In addition, its docile nature and small size make the guinea pig a good species for the examination of allergic responses.

6.3 Guinea Pig Models for Assessing Proteins as Respiratory Allergens

Several guinea pig models have been developed to understand mechanisms of respiratory allergy to protein and many investigators have used ovalbumin as a protein allergen to develop these models (Dunn et al., 1988; Karol et al., 1989; Griffiths-Johnson and Karol, 1991). However, only a few models have been developed to assess the potential of proteins to be occupational allergens; these models were developed with detergent enzymes as the occupational allergens.

6.3.1 Inhalation Model

A general protocol for sensitizing guinea pigs to protein aerosols was developed by Karol and colleagues (1985) and was used successfully to sensitize animals to the detergent enzyme, subtilisin (Thorne et al., 1986; Hillebrand et al., 1987). In this model, animals are exposed to a protein aerosol for 10 or 15 min per day for 5 consecutive days. The animals are subsequently re-exposed to the protein aerosol for 10 min each day starting 5 days after the last exposure. Sera are collected for the measurement of specific antibody. The presence of allergen-specific antibody shows that the animal is immunologically sensitized to the protein. The presence of immediate-onset or late-onset pulmonary symptoms (visual observation of laboured breathing and/or the use of plethysmography to measure changes in breathing rate and pattern) indicate pulmonary sensitization. An elevation of core body temperature (febrile response) can accompany late-onset pulmonary responses (Thorne and Karol, 1989).

Animals exposed to 150 μg of subtilisin protein . m^{-3} for 15 min per day for 5 consecutive days developed immediate-onset and late-onset pulmonary responses when re-exposed to the protein aerosol (Thorne et al., 1986). The immediate-onset pulmonary responses were detected as significant increases in the breathing rate. The late-onset responses were also detected as significant increases in breathing rate accompanied by febrile responses. In addition, the animals possessed allergic antibody specific to subtilisin, although not every animal with antibody developed pulmonary symptoms (Hillebrand et al., 1987). Animals exposed to 8 or 41 μg of subtilisin protein . m^{-3} did not experience pulmonary symptoms, but did develop low levels of antibody to subtilisin. These experiments showed that immune reactivity to this occupational allergen occurred at exposure levels lower than those required for the elicitation of pulmonary responses.

6.3.2 Intratracheal Model

In the guinea pig intratracheal (GPIT) test originally devised for the evaluation of the allergenic potential of detergent enzymes, animals are dosed intratracheally once per week for 10 to 12 consecutive weeks with a protein enzyme allergen (Ritz et al., 1993). Immediately after each dose, animals are observed for signs of immediate-onset pulmonary symptoms. The severity of the symptoms is graded by counting the number of normal respirations that occur between diaphragmatic contractions or spasms. Serum samples are collected at selected time intervals and assessed for allergic antibody to the enzyme allergen. The enzyme subtilisin (Alcalase®) was used as the model allergen in this test since a threshold limit value (TLV) has been established for this protein (American Conference of Governmental Industrial Hygienists, 1990). In addition, a wealth of human data pertaining to antibody and pulmonary responses to this enzyme in the exposed workforce exists (Juniper et al., 1977; Flood et al., 1985; Sarlo et al., 1990) and provides a basis for comparing responses in man to those in the guinea pig.

The allergic antibody dose–response to Alcalase in the GPIT test was comparable with the allergic antibody dose–response to Alcalase in a guinea pig inhalation test, an observation that supports intratracheal instillation as an acceptable alternative to inhalation for delivering protein allergen to the respiratory tract (Ritz et al., 1993). In addition, the development of immediate-onset pulmonary symptoms occurred over the same time-frame (4 to 5 weeks of intratracheal instillation or inhalation exposure). As with the inhalation studies described in Section 6.3.1, animals with pulmonary symptoms had allergic antibody to Alcalase, but not all animals developing allergic antibody experienced pulmonary symptoms. This observation is consistent with the occupational experience in that workers with allergic antibody to enzyme can continue to work in the detergent industry free of pulmonary symptoms (Gaines, 1994). This model has also been used to compare new enzymes and enzyme mixtures with Alcalase for allergenic potency (Sarlo, 1994). Allergenic potency is defined by the dose of enzyme protein needed to induce allergic antibody titres similar to those obtained with selected doses of Alcalase protein. New enzymes can be assessed as more potent than Alcalase, less potent than Alcalase or equipotent to Alcalase. This information has been used to develop operational exposure guidelines to protect workers against developing allergic antibody to new enzymes.

6.3.3 Injection Model

A guinea pig injection model has been developed for comparison of the allergenic potency of new enzymes to that of Alcalase (Blaikie et al., 1994). In this model, guinea pigs receive intradermal injections of enzyme on days 1, 22 and 43 of the test. The amount of enzyme delivered is based on either protein or enzyme activity. Fourteen days after each injection, the presence of serum allergic antibody to the enzyme is assessed by the PCA test and ELISA (for specific IgG antibody). The presence of tissue-fixed allergic antibody is assessed by a modified skin test. Pulmonary symptoms are not measured in this model. The number of animals developing antibody to a new enzyme is compared with the number of animals developing antibody to the reference enzyme Alcalase. The difference in the development of the response rate is used to assign potency to the new enzyme, with this information being used to develop operational exposure guidelines for the manufacturing facilities. The ranking of new enzymes compared with Alcalase with this injection model is very similar to that obtained with the intratracheal model.

6.4 Guinea Pig Models used to Assess Chemicals as Respiratory Allergens

LMW chemicals (MW < 1000) are generally too small to be recognized by the immune system. When these chemicals bind *in vivo* to a larger carrier molecule (usually protein), this chemical–carrier conjugate may be recognized by the immune system. Allergic IgE antibody specific to the chemical or to new antigenic determinants formed by the combination of chemical to carrier molecule can develop and mediate respiratory responsiveness to the chemical. Clinically, not all individuals with pulmonary symptoms to a specific chemical have demonstrable specific allergic antibody to the chemical. This situation is common among workers with occupational asthma to toluene diisocyanate (TDI) (Karol and Jin, 1991; Karol *et al.*, 1994). Therefore, other unidentified immune or non-immune mechanisms may play a role in respiratory responses to chemicals.

Several of the guinea pig models developed to assess chemicals as respiratory allergens rely on chemical–protein conjugates as antigens for measuring chemical-specific allergic antibody. Human serum albumin is often used as a carrier protein for developing antigens that can be used as reagents to test for the presence of antibody in chemically exposed humans. In guinea pig models, the homologous protein, guinea pig serum albumin (GSA), is the common carrier protein used to develop the antigenic reagent (Sarlo and Clark, 1992; Sarlo and Karol, 1994). Work done by Jin *et al.* (1993) showed that GSA was one of several proteins in bronchoalveolar lavage fluid adducted with TDI after animals were exposed by inhalation to TDI atmospheres. Careful characterization of chemical–GSA conjugates is important for the assessment process and, as yet, there are no standard methods for preparing such conjugates. However, new analytical techniques in chromotography and mass spectroscopy help in the characterization of conjugates used for the assessment of chemicals as allergens (Gauggel *et al.*, 1993; Asquith *et al.*, 1994).

Since non-immune mechanisms may play a role in occupational allergy to certain chemicals, some of the animal models also rely on pulmonary symptoms as a measure of reactivity to the chemical. Symptoms may be elicited from these animals by exposure to atmospheres of free chemical or chemical–carrier conjugate. In certain cases, pulmonary sensitivity could only be demonstrated by exposure to chemical–carrier conjugates and not to the chemical alone (Sarlo and Karol, 1994).

6.4.1 Inhalation Model

A guinea pig inhalation model was developed by Karol (1983) to examine immune and pulmonary responses to TDI. Since that time, this model has been used by other investigators to examine immune and pulmonary responses to TDI and other chemical respiratory allergens such as diphenylmethane diisocyanate (MDI) (Karol and Thorne, 1988), trimeric hexamethylene diisocyanate (Des-N) (Pauluhn and Eben, 1991), trimellitic anhydride (TMA) (Botham et al., 1988; Pauluhn and Eben, 1991), phthalic anhydride (PA) (Sarlo and Clark, 1992; Sarlo et al., 1994), procion yellow MX4R (Botham et al., 1988) and reactive black B dye (Sarlo and Clark, 1992). The investigators were able to reproduce Karol's TDI experience, an observation which points to the reproducibility of the model. In this model, animals are housed in whole-body glass plethysmographs and exposed for 3 h per day for 5 consecutive days to a chemical atmosphere. Sixteen days after the last exposure, animals are exposed to aerosols of chemical–carrier conjugate and pulmonary function is continuously monitored over a 24-h period to detect immediate-onset and late-onset pulmonary responses. Serum samples are collected to look for chemical-specific allergic antibody.

The results of the various studies have been summarized (Sarlo and Karol, 1994). Briefly, pulmonary reactivity was detected in those animals exposed to the isocyanates and the anhydrides (with the exception of Des-N) and challenged with the appropriate chemical–protein conjugate. Pulmonary responses were also elicited by challenge with free chemical from animals exposed to TDI, MDI and TMA. No pulmonary reactivity was elicited from animals exposed to the known respiratory allergens black B dye or the procion yellow dye and challenged with either dye–protein conjugate or free dye. Animals sensitized and challenged with phthalic anhydride displayed lung pathology that was consistent with the haemorrhagic responses sometimes observed with TMA allergy (Sarlo et al., 1994).

Specific allergic IgG1 antibody (measured by PCA) and specific IgG antibody (measured by ELISA) were found for each chemical. Generally, as the exposure concentration of the chemical was increased, the amount of antibody produced to the chemical increased. In addition, the number of animals responding with allergic antibody production increased with increased exposure. Animals with pulmonary reactivity had antibody to the chemical. However, not all animals demonstrating allergic antibody experienced pulmonary responses. This observation is similar to those made with protein allergens in the intratracheal and inhalation models.

The exposure conditions needed to induce an immune response may be different from those required to elicit pulmonary reactions. In one inhalation study of TDI, the investigators found that animals exposed to 0.02 ppm TDI did not make antibody to this chemical but animals exposed to concentrations of greater than 0.2 ppm did make TDI-specific antibody (Aoyama et al., 1994). When challenged with 0.02 ppm TDI, animals with antibody to the chemical experienced immediate-onset pulmonary symptoms. There was no relationship between antibody titre and the pulmonary response. This study showed that pulmonary symptoms could be elicited by an exposure concentration that does not induce antibody production.

6.4.2 Injection Models

Several guinea pig injection models have been developed as an approach to assess quickly and cheaply whether or not a chemical is immunogenic and allergenic. Some of the models were also developed as a way to sensitize animals quickly in order to assess pulmonary reactions upon inhalation challenge with chemical. In two of these models, animals receive one or three intradermal injections of the highest tolerated dose of chemical (Botham et al., 1989; Pauluhn and Eben, 1991). Two weeks later, sera are collected and the animals are challenged, via inhalation, with free chemical or chemical-protein conjugate. The investigators were able to generate antibody to TMA or Des-N with this protocol. Only animals challenged with high concentrations of TMA dust (approximately 45 mg . m^{-3}) experienced pulmonary responses. Animals exposed to atmospheres of Des-N did not exhibit pulmonary reactions. In a similar study, animals received a single intradermal injection of 0.0003 to 0.3 per cent MDI (Rattray et al., 1994). IgG1 antibody to MDI was detected by PCA and ELISA; IgE was also found in the 4-day PCA test. An MDI-GSA conjugate was used as the antigen in the serological assays. Inhalation challenge with 27 to 36 mg . m^{-3} MDI resulted in immediate-onset pulmonary responses in some of the animals. No antibody or pulmonary reactivity was noted in the animals injected with 0.0003 per cent MDI. There was a dose-dependent increase in antibody titre and the number of animals showing pulmonary reactivity with increasing induction doses of MDI (0.003 per cent and higher).

A two-centre study was recently conducted to examine the reproducibility of the intradermal injection model (Blaikie et al., 1995). Animals were injected with the highest tolerated dose of TMA, MDI, PA or TDI. The contact allergen, 2,4-dinitrochlorobenzene (DNCB) was tested as a control chemical. After 3 weeks, serum samples were collected for antibody measurements and the animals were challenged, via inhalation, with free chemical. Both centres found allergic antibody to each chemical (except DNCB) by the PCA test and the ELISA using appropriate chemical-GSA conjugates as antigens. There were very low IgG antibody titres to DNCB, as determined by ELISA. Both laboratories were able to elicit immediate-onset pulmonary responses from animals exposed to relatively high concentrations of TMA (5 to 50 mg . m^{-3}), MDI (18 to 55 mg . m^{-3}) and PA (11 to 52 mg . m^{-3}). No pulmonary reactions were elicited with TDI (0.13 to 2 ppm). However, the TDI-exposed animals did experience pulmonary reactions when challenged with a TDI-GSA conjugate. In contrast, there were no pulmonary responses observed among the DNCB-exposed animals. This study showed that a single injection of very high level of free chemical can sensitize animals and that very high atmospheric concentrations of free chemical can elicit pulmonary responses in these animals. How this information can be used to protect workers from developing pulmonary symptoms is unclear at this time.

Another injection model utilizes multiple subcutaneous injections of free chemical over several weeks (Sarlo and Clark, 1992). Animals are injected with several molar doses of chemical allergen twice per week for 4 consecutive weeks, followed by 1 week of non-treatment, followed by two additional booster injections. Seven days after the last injection, animals are assessed for specific circulating antibody to the chemical by PCA or ELISA and for tissue-fixed allergic antibody by the active cutaneous anaphylaxis test. Animals can also be examined for immediate-onset pulmonary reactivity to the chemical by intratracheal instillation of chemical–protein conjugate. Six respiratory chemical allergens have been tested in this model: TMA, PA, TDI, MDI, reactive black B dye and hexachloroplatinic acid (Sarlo and Clark, 1992; Sarlo and Karol, 1994). Since animals are injected with three or four doses of chemical over time, dose–response relationships can be constructed. Also, since chemicals are delivered on an equimolar basis (rather than weight basis), comparisons of the antibody responses and/or respiratory responses can be made among the chemicals. For example, the greatest antibody responses were noted among animals injected with TMA, PA and TDI. Animals injected with MDI and the dye did not generate as vigorous an antibody response. It has been suggested that these types of comparisons can be used to compare potencies among chemicals.

6.4.3 Topical Models

Occupational exposure to chemical allergens can occur through the respiratory tract, mucous membranes and the skin. Dermal exposure to reactive chemicals may play a role in the development of allergic antibody to that chemical. This premise is the basis of a mouse model that examines alterations of total serum IgE levels after topical application of respiratory chemical allergens (Dearman et al., 1992; Dearman and Kimber, 1992). Several studies have examined antibody and pulmonary responses of guinea pigs after dermal exposure to various isocyanates. One or two dermal applications of 25 or 100 per cent TDI led to the generation of TDI-specific allergic antibody and one-third of the animals experienced pulmonary reactions upon inhalation exposure to TDI or TDI–GSA conjugate (Karol et al., 1981a). In contrast, dermal application of 1 to 100 per cent dicyclohexylmethane-4,4'-diisocyanate (HMDI) did not result in antibody production or pulmonary reactivity (Karol and Magreni, 1982). These animals did develop contact sensitization to HMDI, an observation consistent with the clinical history of HMDI. When animals were exposed to a single, 6-hour occluded application of 10, 30 or 100 per cent MDI, a dose-dependent increase in the number of animals with antibody to MDI was noted (Rattray et al., 1994). In addition, pulmonary reactivity was noted in about a quarter of the animals exposed to atmospheres containing between 25.9 and 36.4 mg $\cdot$ m^{-3} MDI. Investigators have had mixed success with inducing antibody and pulmonary responses in animals exposed to MDI via the inhalation route (compare Karol and Thorne, 1988 with Rattray et al., 1994). It is possible that for certain chemicals, dermal exposure can be a relevant route of exposure for sensitization of the respiratory tract.

6.4.4 Miscellaneous Models

Other variations of injection models have been used by several investigators to examine antibody, pulmonary and neurogenic responses to chemical allergens. In one model, employed by Tao et al. (1991), guinea pigs received an intramuscular injection of TMA conjugated to carrier protein. Several weeks after the injection, animals were exposed to 130 mg $\cdot$ m^{-3} TMA dust. These animals experienced significant changes in lung pathology (most notably haemorrhage) consistent with one of the TMA syndromes described by Zeiss et al. (1977, 1983). The same investigators also found that these animals produced both IgG1 and IgE antibody to TMA and developed airway hyperreactivity accompanied by eosinophilic influx (Obata et al., 1992).

In a similar model used by Arakawa et al. (1993) and Hayes et al. (1995), animals injected with TMA and challenged by intratracheal instillation of TMA–GSA conjugate experienced a significant increase in airway resistance and significant extravasation of Evans blue dye into the airways. The IgG1 antibody titre did not correlate with resistance changes, but did correlate with extravasation. Treatment of the animals with an antihistamine alleviated changes in airway resistance, but had no effect on extravasation. Rather, treatment of animals with capsaicin and hexamethonium (to block release of neurogenic mediators) alleviated the extravasation, but had no effect on airway resistance. These findings suggest that chemical allergens, like TMA, may mediate pulmonary reactions by different mechanisms and antibody responses may correlate positively with some of these mechanisms.

6.5 The Use of Guinea Pig Models in Risk Assessment

The various guinea pig models developed to assess proteins or chemicals as respiratory allergens may be used in a risk assessment process. One can assume that exposure to any foreign protein (e.g. bacterial/fungal enzymes, plant proteins, animal dander) can lead to recognition by the immune system and induce the production of allergic IgE antibody in some individuals. Therefore, the risk assessment process revolves around how to protect workers from developing antibody and, eventually, symptoms to these proteins. The situation with chemicals is different. Not all reactive chemicals have the capacity to induce allergic respiratory reactions (e.g. contact allergens such as DNCB or oxazolone). Therefore, the assessment includes identification of the chemical as an allergen and the development of plans to protect workers from developing allergy once exposed to the chemical.

In the detergent enzyme industry, new enzymes are evaluated in the GPIT test or the injection test (Sarlo, 1994). Since a TLV exists for the protease, Alcalase, and much human data exist pertaining to antibody and respiratory responses to this enzyme, one can compare new enzymes with Alcalase in the guinea pig models. By comparing antibody and/or respiratory responses to new enzymes to responses obtained with Alcalase, new enzymes can be judged as more potent, less potent or equivalent to Alcalase. The potency differences can be used to establish TLVs and other exposure guidelines for the new enzymes. Evaluation of sensitization among workers exposed to these enzymes suggests that these exposure guidelines are effective in protecting workers.

Assessment of chemicals as allergens is more difficult. The majority of guinea pig models are based on the assumption that respiratory allergy and asthma to chemicals is mediated through an allergic antibody response. Therefore, finding antibody to a chemical after injection or inhalation exposure is one approach to identifying the chemical as a potential allergen. However, respiratory symptoms due to chemical exposure may be mediated by other, as yet, unidentified mechanisms. Measuring pulmonary reactivity of animals after chemical exposure is one approach to assess these 'mechanisms'. This approach is not favoured for routine risk assessment, since inhalation testing is expensive and time-consuming and investigators have had mixed success with eliciting pulmonary reactivity in animals after exposure to free chemical. Also, an antibody response may be the only response obtained in the animal model. This was observed with the known occupational allergen, reactive black B dye. Therefore, measuring antibody as a surrogate marker is a current best approach for assessing chemicals as allergens (Selgrade et al., 1994). Antibody measurements are not without problems, since these measurements are only as good as the chemical–protein conjugate used in the serological tests.

It has been suggested that one can compare the potential of a chemical to be an allergen to known chemical allergens in the animal models (Sarlo and Clark, 1992; Briatico-Vangosa et al., 1994; Sarlo and Karol, 1994; Selgrade et al., 1994). This approach is similar to the one used by the detergent enzyme industry. A benchmark chemical, such as TMA, can be used for comparative purposes in these models. TMA may be an ideal benchmark since data exist to support the TLV for TMA. For example, epidemiological studies have shown that new sensitizations and the occurrence of asthma are uncommon events in those manufacturing sites that adhere to the exposure guidelines (Zeiss et al., 1990). Therefore, by comparing antibody and/or pulmonary responses of animals to a new chemical with those responses obtained with TMA, one can use the similarities or differences between the two chemicals to help establish exposure guidelines. Confirmation of the adequacy of these guidelines would require extensive monitoring of the exposed workforce.

References

AMERICAN CONFERENCE OF GOVERNMENTAL INDUSTRIAL HYGIENISTS (1990) In *Threshold Limit Values and Biological Exposure Indices*, p. 38, Cincinnati.

AOYAMA, K., HUANG, J., UEDA, A. & MATSUSHITA, T. (1994) Provocation of respiratory allergy in guinea pigs following inhalation of free toluene diisocyanate. *Archives of Environmental Contamination and Toxicology*, 26, 403–407.

ARAKAWA, H., LOTVALL, J., KAWIKOVA, I., TEE, R., HAYES, J., LOFDAHL, C. G., NEWMAN TAYLOR, A. J. & SKOOGH, B. E. (1993) Airway allergy to trimellitic anhydride in guinea pigs: different time courses of IgG1 titre and airway responses to allergen challenge. *Journal of Allergy and Clinical Immunology*, 92, 425–434.

ASQUITH, T. N., KEOUGH, T. W., GAUGGEL, D. L., TAKIGIKU, R., SARLO, K. & LACEY, M. P. (1994) Characterization by mass spectroscopy of conjugates between proteins and low molecular weight chemicals, in CRABB, J. W. (Ed.) *Techniques in Protein Chemistry V*, pp. 11–18. New York: Academic Press.

BLAIKIE, L., BASKETTER, D. A. & MORROW, T. (1994) Experience with a guinea pig model for the assessment of respiratory allergens. *Human and Experimental Toxicology*, 9, 743.

BLAIKIE, L., MORROW, T., WILSON, A. P., HEXT, P., HARTOP, P. J., RATTRAY, N. J., WOODCOCK, D. & BOTHAM, P. A. (1995) A two-centre study for the evaluation and validation of an animal model for the assessment of the potential of small molecular weight chemicals to cause respiratory allergy. *Toxicology*, 96, 37–50.

BOTHAM, P. A., HEXT, P. M., RATTRAY, N. J., WALSH, S. T. & WOODCOCK, D. R. (1988) Sensitization of guinea pigs by inhalation exposure to low molecular weight chemicals. *Toxicology Letters*, 41, 159–173.

BOTHAM, P. A., RATTRAY, N. J., WOODCOCK, D. R., WALSH, S. T. & HEXT, P. M. (1989) The induction of respiratory allergy in guinea pigs following intradermal injection of TMA: a comparison with the response to DNCB. *Toxicology Letters*, 47, 25–39.

BRIATICO-VANGOSA, G., BRAUN, C. L. J., COOKMAN, G. R., HOFFMAN, T., KIMBER, I., LOVELESS, S. E., MORROW, T., PAULUHN, J., SORENSEN, T. & NIESSEN, H. J. (1994) Respiratory allergy: hazard identification and risk assessment. *Fundamental and Applied Toxicology*, 23, 145–158.

CHAN-YEUNG, M. (1990) Occupational asthma. *Chest*, 98, 148S–161S.

DEARMAN, R. J. & KIMBER, I. (1992) Divergent immune responses to respiratory and contact chemical allergens: antibody elicited by phthalic anhydride and oxazolone. *Clinical and Experimental Allergy*, 22, 241–250.

DEARMAN, R. J., BASKETTER, D. A. & KIMBER, I. (1992) Variable effects of chemical allergens on serum IgE concentrations in mice: preliminary evaluation of a novel approach to the identification of respiratory sensitizers. *Journal of Applied Toxicology*, 12, 317–323.

DOBSON, C., MORSETH, D. J. & SOULSBY, J. L. (1971) Immunoglobulin E-type antibodies induced by *Ascaris suum* infections in guinea pigs. *Journal of Immunology*, 106, 128–133.

DUNN, C. J., ELLIOTT, G. A., OOSTVEEN, J. A. & RICHARDS, I. M. (1988) Development of a prolonged eosinophil-rich inflammatory leukocyte infiltration in the guinea pig asthmatic response to ovalbumin inhalation. *American Review of Respiratory Disease*, 137, 541–547.

FLOOD, D. F. S., BLOFELD, R. E., BRUCE, C. F., HEWITT, J. I., JUNIPER, C. P. & ROBERTS, D. M. (1985) Lung function, atopy, specific hypersensitivity and smoking of workers in the enzyme detergent industry over 11 years. *British Journal of Industrial Medicine*, **42**, 43–50.

GAINES, W. G. (1994) Occupational health experience manufacturing multiple enzyme detergents and methods to control enzyme exposures. *Proceedings of the Toxicology Forum*, 143–147.

GAUGGEL, D. L., SARLO, K. & ASQUITH, T. N. (1993) A proposed screen for evaluating low molecular weight chemicals as potential respiratory allergens. *Journal of Applied Toxicology*, **13**, 307–313.

GRIFFITH-JOHNSON, D. A. & KAROL, M. H. (1991) Validation of a non-invasive technique to assess development of airway hyperreactivity in animal models of immunologic pulmonary hypersensitivity. *Toxicology*, **65**, 283–294.

GRIFFITH-JOHNSON, D. A., JIN, R. & KAROL, M. H. (1993) The role of purified IgG1 in pulmonary hypersensitivity responses of the guinea pig. *Journal of Toxicology and Environmental Health*, **40**, 117–127.

HAYES, J. P., HAN-PIN, K., ROHDE, J. A. L., NEWMAN TAYLOR, A. J., BARNES, P. J., FAN-CHUNG, K. & ROGERS, D. F. (1995) Neurogenic goblet cell secretion and bronchoconstriction in guinea pigs sensitized to trimellitic anhydride. *European Journal of Pharmacology, Environmental Toxicology and Pharmacology Section*, **292**, 127–134.

HILLEBRAND, J., THORNE, P. T. & KAROL, M. H. (1987) Experimental sensitization to subtilisin: production of specific antibodies following inhalation exposure of guinea pigs. *Toxicology and Applied Pharmacology*, **89**, 449–456.

JIN, R., DAY, B. W. & KAROL, M. H. (1993) TDI protein adducts in the bronchoalveolar lavage of guinea pigs exposed to vapors of the chemical. *Chemical Research Toxicology*, **6**, 906–912.

JUNIPER, C. P., HOW, M. J., GOODWIN, B. F. J. & KINSHOTT, A. K. (1977) *Bacillus subtilis* enzymes: a seven year clinical epidemiological and immunological study of an industrial allergen. *Journal of the Society of Occupational Medicine*, **27**, 3–12.

KAROL, M. H. (1983) Concentration dependent immunologic response to TDI following inhalation exposure. *Toxicology and Applied Pharmacology*, **68**, 229–241.

KAROL, M. H. & JIN, R. (1991) Mechanisms of immunotoxicity to isocyanates. *Chemical Research in Toxicology*, **4**, 503–509.

KAROL, M. H. & MAGRENI, C. M. (1982) Extensive skin sensitization with minimal antibody production in guinea pigs as a result of exposure to HMDI. *Toxicology and Applied Pharmacology*, **65**, 291–301.

KAROL, M. H. & THORNE, P. S. (1988) Pulmonary hypersensitivity and hyperreactivity: implications for assessing allergic responses, in GARDNER, D. E., CRAPO, J. D. & MASSARO, E. J. (Eds) *Toxicology of the Lung*, pp. 427–448. New York: Raven Press.

KAROL, M. H., HAUTH, B. A., RILEY, E. J. & MAGRENI, C. M. (1981a) Dermal contact with TDI produces respiratory tract hypersensitivity in guinea pigs. *Toxicology and Applied Pharmacology*, **58**, 221–230.

KAROL, M. H., STADLER, J., UNDERHILL, D. & ALARIE, Y. (1981b) Monitoring delayed onset pulmonary hypersensitivity in guinea pigs. *Toxicology and Applied Pharmacology*, **61**, 277–285.

KAROL, M. H., STADLER, J. & MAGRENI, C. M. (1985) Immunotoxicologic evaluation of the respiratory system: animal models for immediate and delayed-onset pulmonary hypersensitivity. *Fundamental and Applied Toxicology*, **5**, 459–472.

KAROL, M. H., HILLEBRAND, J. & THORNE, P. T. (1989) Characteristics of weekly pulmonary hypersensitivity responses elicited in guinea pigs by inhalation of ovalbumin aerosols. *Toxicology and Applied Pharmacology*, **100**, 234–246.

KAROL, M. H., TOLLERUD, D. J., CAMPBELL, T. P., FABBRI, L., MAESTRELLI, P., SAETTA, M. & MAPP, C. E. (1994) Predictive value of airway hyperresponsiveness and circulating IgE for identifying types of responses to TDI inhalation challenge. *American Journal of Respiratory Critical Care Medicine*, **149**, 611–615.

KAWABATA, T. T., BABCOCK, L. S., GAUGGEL, D. L., ASQUITH, T. N., FLETCHER, E. R., HORN, P. A., RATAJCZAK, H. V. & GRAZIANO, F. M. (1995) Optimization and validation of an ELISA to measure specific guinea pig IgG1 antibody as an alternative to the *in vivo* PCA assay. *Fundamental and Applied Toxicology*, **24**, 238–246.

LAPA E SILVA, J. R., BACHELET, C. M., PRETOLANI, M., BAKER, D., SCHEPER, R. J. & VARGAFTIG, B. B. (1993) Immunopathologic alterations in the bronchi of immunized guinea pigs. *American Journal of Respiratory Cell and Molecular Biology*, **9**, 44–53.

LAPA E SILVA, J. R., PRETOLANI, M., BACHELET, C. M., BAKER, D., SCHEPER, R. J. & VARGAFTIG, B. B. (1994) Emergence of T lymphocytes, eosinophils and dendritic cells in the bronchi of actively sensitized guinea pigs after antigenic challenge. *Brazilian Journal of Medical Biological Research*, **27**, 1653–1658.

OBATA, H., TAO, Y., KIDO, M., NAGATA, N., TANAKA, I. & KUROIWA, A. (1992) Guinea pig model of immunologic asthma induced by inhalation of TMA. *American Review of Respiratory Disease*, **146**, 1553–1558.

VAN OOSTERHOUT, A. J. M., VAN ARK, I., FOLKERTS, G., VAN DER LINDE, H. J., SAVELKOUL, H. F. J., VERHEYEN, A. C. K. P. & NIJKAMP, F. P. (1995) Antibody to interleukin 5 inhibits virus induced airway hyperresponsiveness to histamine in guinea pigs. *American Journal of Respiratory Critical Care Medicine*, **151**, 177–183.

PARISH, W. E. (1972) Eosinophilia: cutaneous eosinophilia in guinea pigs mediated by passive anaphylaxis with IgG1 or reagin, and antigen–antibody complexes: its relation to neutrophils and mast cells. *Immunology*, **23**, 19–34.

PAULUHN, J. & EBEN, A. (1991) Validation of a non-invasive technique to assess immediate or delayed onset airway hypersensitivity in guinea pigs. *Journal of Applied Toxicology*, **11**, 423–431.

PAUPE, J. (1984) *L'allergie*. Paris: Presses Universitaire de France.

PRENTO, A. (1991) ELISA for guinea pig IgG1: an alternative to the PCA test for relative allergenicity. *Alternatives to Laboratory Animals*, **19**, 8–14.

PRETOLANI, M., RUFFIE, C., LAPA E SILVA, J. R., JOSEPH, D., LOBB, R. R. & VARGAFTIG, B. B. (1994) Antibody to very late activation antigen 4 prevents antigen induced bronchial hyperreactivity and cellular infiltration in the guinea pig airways. *Journal of Experimental Medicine*, **180**, 795–805.

RATNER, B., JACKSON, H. C. & GRUEHL, H. L. (1927) Respiratory anaphylaxis: sensitization, shock, bronchial asthma and death induced in the guinea pig by nasal inhalation of dry horse dander. *American Journal of Diseases of the Child*, **34**, 23–52.

Rattray, N. J., Botham, P. A., Hext, P. M., Woodcock, D. R., Fielding, I., Dearman, R. J. & Kimber, I. (1994) Induction of respiratory hypersensitivity to MDI in guinea pigs: influence of route of exposure. *Toxicology*, **88**, 15–30.

Ritz, H. L., Evans, B. L. B., Bruce, R. D., Fletcher, E. R., Fisher, G. L. & Sarlo, K. (1993) Respiratory and immunological responses of guinea pigs to enzyme containing detergents: a comparison of intratracheal and inhalation modes of exposure. *Fundamental and Applied Toxicology*, **21**, 31–37.

Sarlo, K. (1994) Human health risk assessment: focus on enzymes, in Cahn, A. (Ed.) *Proceedings of the Third World Conference on Detergents, a Global Perspective*, pp. 54–57. Champaigne, Ill.: AOCS Press.

Sarlo, K. & Clark, E. D. (1992) A tier approach for evaluating the respiratory allergenicity of low molecular weight chemicals. *Fundamental and Applied Toxicology*, **18**, 107–114.

Sarlo, K. & Karol, M. H. (1994) Guinea pig predictive tests for respiratory allergy, in Dean, J. H., Luster, M. I., Munson, A. E. & Kimber, I. (Eds) *Immunotoxicology and Immunopharmacology*, pp. 703–720. New York: Raven Press.

Sarlo, K., Clark, E. D., Ryan, C. R. & Bernstein, D. I. (1990) ELISA for human IgE antibody to subtilisin A (Alcalase): correlation with RAST and skin test results with occupationally exposed individuals. *Journal of Allergy and Clinical Immunology*, **86**, 393–399.

Sarlo, K., Clark, E. D., Ferguson, J., Zeiss, C. R. & Hatoum, N. (1994) Induction of type I hypersensitivity in guinea pigs after inhalation of phthalic anhydride. *Journal of Allergy and Clinical Immunology*, **94**, 747–756.

Selgrade, M. K., Zeiss, C. R., Karol, M. H., Sarlo, K., Kimber, I., Tepper, J. S. & Henry, M. C. (1994) Workshop on the status of test methods for assessing potential mechanisms of chemicals to induce respiratory allergic reactions. *Inhalation Toxicology*, **6**, 303–319.

Tao. Y., Sugiura, T., Nakamura, H., Kido, M., Tanaka, I. & Kuriona, A. (1991) Experimental lung injury induced by TMA inhalation in guinea pigs. *International Archives of Allergy and Applied Immunology*, **96**, 119–127.

Thorne, P. T. & Karol, M. H. (1989) Association of fever with late onset pulmonary hypersensitivity responses in the guinea pig. *Toxicology and Applied Pharmacology*, **100**, 247–258.

Thorne, P. T., Hillebrand, J., Magreni, C., Riley, E. J. & Karol, M. H. (1986) Experimental sensitization to subtilisin: production of immediate and late onset pulmonary reactions. *Toxicology and Applied Pharmacology*, **86**, 112–123.

Zeiss, C. R., Patterson, R., Pruzansky, J. J., Miller, M., Rosenberg, M. & Levitz, D. (1977) TMA induced airway syndromes: clinical and immunological studies. *Journal of Allergy and Clinical Immunology*, **60**, 96–103.

Zeiss, C. R., Wolkonsky, P., Chacon, R., Tuntland, A. P., Levitz, D., Pruzansky, J. J. & Patterson, R. (1983) Syndromes in workers exposed to TMA: a longitudinal clinical and immunological study. *Annals of Internal Medicine*, **98** 8–12.

Zeiss, C. R., Mitchell, J. H., van Peener, P. F. D., Harris, J. & Levitz, D. (1990) A 12 year clinical and immunologic evaluation of workers involved in the manufacture of TMA. *Allergy Proceedings*, **11**, 71–72.

7

Predictive Assessment of Respiratory Sensitizing Potential of Chemicals in Mice

IAN KIMBER and REBECCA J. DEARMAN

Zeneca Central Toxicology Laboratory, Alderley Park, Macclesfield

7.1 Introduction

As described in the previous chapter, the guinea pig has conventionally been the species of choice for toxicological assessment of chemical respiratory allergy. The question this chapter addresses is whether there exist realistic opportunities to evaluate in mice the ability of chemicals to cause sensitization of the respiratory tract. An important consideration in the use of guinea pigs in this respect has been the fact that challenge-induced pulmonary reactions can be elicited in this species that resemble in some ways human allergic asthma. It is less easy, although by no means impossible (Garssen *et al.*, 1991), to measure accurately allergen-induced respiratory reactions in mice and for this reason initiatives using this species have focused largely upon characterization of immune responses by chemical sensitizers.

The term allergy describes the state of heightened responsiveness and associated adverse health effects that result from the stimulation (by protein or chemical allergens) of a specific immune response. By definition, therefore, chemical allergens have the potential to induce in susceptible individuals the quantity and quality of immune response necessary to effect allergic sensitization such that, following second or subsequent exposure to the inducing allergen, a hypersensitivity reaction will be provoked. An important research objective has been to define in mice the quality of immune response stimulated by chemicals that are known in humans to cause respiratory allergy and occupational asthma. It is from such investigations that opportunities have arisen to identify potential chemical respiratory allergens as a function of the characteristics of the immune response they elicit.

7.2 Immunological Background

Chemicals may cause various types of allergic disease, those of greatest importance in the context of occupational health being contact sensitization (allergic contact dermatitis) and respiratory hypersensitivity. The intriguing question is why some chemical allergens are associated in humans primarily with contact sensitization, whereas others are considered preferentially to elicit respiratory allergy (Kimber, 1995). Contact sensitization results from the stimulation of a specific T-lymphocyte response and it is the action of effector T cells that initiates the cutaneous inflammatory response that is recognized clinically as allergic contact dermatitis. Although respiratory allergy induced by proteins (such as pollen and house dust mite allergens) is regarded usually as being effected by IgE antibody-associated mechanisms, there is less certainty with respect to pulmonary hypersensitivity caused by chemicals. Such uncertainty arises from the fact that in many instances, particularly in the case of occupational asthma induced by isocyanates, it has not been possible to demonstrate the presence of specific IgE antibodies in some patients who are clearly symptomatic (Chan-Yeung, 1995). Nevertheless, there have been reported associations between respiratory hypersensitivity to chemical allergens and IgE antibody (Moller et al., 1985; Salvaggio et al., 1988; Drexler et al., 1993; Karol et al., 1994; Nielsen et al., 1994). It may prove that, in some instances, the failure to detect specific IgE antibody is due to the use of techniques that are inappropriate or of insufficient sensitivity. If such is the case then the association between chemical respiratory allergy and IgE may be somewhat greater than appreciated currently. However, it has to be borne in mind that immune mechanisms independent of IgE antibody may, in some instances and with some chemical allergens, play an important pathogenetic role in the elicitation of respiratory allergy, particularly in the development of less acute reactions (Fabbri et al., 1994; Kimber, 1996).

Notwithstanding these considerations, there is clear evidence that chemicals regarded as being primarily either contact or respiratory allergens in humans provoke in mice qualitatively distinct immune responses (Dearman et al., 1992a; Kimber and Dearman, 1992, 1993, 1994). Results of experiments performed with trimellitic anhydride (TMA) and 2,4-dinitrochlorobenzene (DNCB) illustrate the point. The ability of TMA to cause respiratory allergy and occupational asthma in a proportion of those exposed is well established (Bernstein et al., 1982; Zeiss and Patterson, 1993). In contrast, however, although it is a potent contact allergen, DNCB apparently lacks the potential to cause respiratory sensitization (Botham et al., 1989). It was found that inhalation exposure of mice to atmospheres of TMA or DNCB in both cases resulted in the stimulation of a specific IgG anti-hapten antibody response, but that only with TMA was there the appearance of a specific IgE antibody response also (Dearman et al., 1991). In subsequent investigations it was shown that TMA and DNCB also elicit immune responses of different types when applied topically to mice. Under conditions of exposure where both chemicals caused the activation of draining lymph nodes, comparable levels of draining lymph node cell proliferation and IgG anti-hapten antibody responses of equivalent magnitude, only TMA stimulated the production of IgE antibody (Dearman and Kimber, 1991). It was demonstrated also that TMA and DNCB provoked different isotype patterns of IgG antibody, with the latter chemical inducing a stronger IgG2a antibody response. Moreover, only following exposure of mice to TMA was an increase in the serum concentration of IgE observed (Dearman and Kimber, 1991). That TMA induced in mice a specific IgE antibody response of sufficient vigour to cause systemic sensitization was demonstrated by two experimental observations. First, it was shown that in mice sensitized previously with TMA, dermal challenge with the same chemical resulted in both immediate (1 h) and delayed-type (maximal at 24 h) cutaneous hypersensitivity reactions (Dearman et al., 1992c; Lauerma et al., 1995). Under the same conditions challenge of mice with DNCB caused only a delayed-type reaction (Dearman et al., 1992c). That the immediate reactions observed in TMA-sensitized mice were attributable to IgE antibody was suggested by the fact that the same responses could be transferred to naive recipient mice with serum from animals sensitized with TMA. Furthermore, immediate hypersensitivity to DNCB could be induced in mice by passive intradermal sensitization with a monoclonal IgE anti-dinitrophenol (DNP) antibody (Dearman et al., 1992c). The second line of evidence is supplied by the demonstration that topical exposure of mice to TMA caused the sensitization of peritoneal mast cells in vivo, as measured by the specific conjugate-induced release of 5-hydroxytryptamine in vitro (Holliday et al., 1993).

The qualities of immune responses stimulated by DNCB and TMA were considered to be indicative of the selective activation of distinct populations of T-helper (Th) cells, with DNCB and TMA eliciting, respectively, preferential Th1 cell-type and Th2 cell-type responses. Similarly divergent immune responses were observed following the topical exposure of mice to other known or suspected chemical contact and respiratory allergens and in each case only those chemicals believed to cause sensitization of the respiratory tract were found to stimulate an increase in the serum concentration of IgE (Dearman and Kimber, 1992; Dearman et al., 1992d). The association between the potential to cause respiratory allergy in humans and the ability to induce in mice an increase in serum IgE concentration forms the basis of the mouse IgE test described below.

Functional subpopulations of Th cells are distinguished by discrete cytokine secretion profiles. While both populations in mice produce granulocyte/macrophage colony-stimulating factor (GM-CSF) and interleukin-3 (IL-3), only Th1 cells secrete interferon-γ (IFN-γ) and interleukin-2 (IL-2) and only Th2 cells produce interleukins-4, 5, 6 and 10 (IL-4, IL-5, IL-6 and IL-10) (Mosmann and Coffman, 1989; Mosmann et al., 1991). Consistent with the proposal that chemical contact and respiratory allergens provoke in mice immune responses characteristic of selective Th1- and Th2-type activation, respectively, it has been found recently that draining lymph node cells isolated from sensitized animals exhibit discrete cytokine secretion profiles. Secondary or chronic exposure of mice to chemical respiratory allergens such as TMA or toluene diisocyanate (TDI) was found to induce the production by draining lymph node cells of comparatively high levels of IL-10 and mitogen-inducible IL-4, but only relatively low concentrations of IFN-γ. In contrast, contact allergens such as oxazolone and 2,4-dinitrofluorobenzene (DNFB) that are suspected not to cause respiratory sensitization induced the converse pattern with vigorous IFN-γ production, but comparatively little IL-4 or IL-10 (Dearman et al., 1994, 1995, 1996). These data raise the possibility that cytokine production profiles may provide a means of characterizing the type of allergic response a chemical sensitizer will elicit preferentially. The potential value of 'cytokine fingerprinting' as an approach to the identification and classification of chemical allergens is considered later.

7.3 The Mouse IgE Test

In initial studies it was found that topical exposure of BALB/c strain mice to known chemical respiratory allergens, but not to contact allergens, induced an increase in the serum concentration of IgE (Dearman et al., 1992b). In those investigations, groups of mice received 50 μl of a single concentration of the test chemical on each shaved flank. Seven days later 25 μl of the same solution diluted 1:1 with vehicle were applied to the dorsum of both ears. At various periods following exposure mice were exsanguinated by cardiac puncture and serum prepared. The concentration of IgE in serum samples was measured using a sandwich enzyme-linked immunosorbent assay (ELISA) calibrated using a monoclonal mouse IgE antibody. Results were recorded as serum IgE concentration in $\mu g \cdot ml^{-1}$ and induced changes measured relative to historical control values for the IgE content of serum drawn from naive mice (Dearman et al., 1992b). On the basis of those preliminary studies the decision was made in subsequent investigations to measure the serum concentration of IgE 14 days following initiation of exposure to the test material. Supplementary studies were conducted to examine dose–response relationships in the mouse IgE test. It was found that each of four chemical respiratory allergens, TMA, TDI, diphenylmethane diisocyanate (MDI) and hexamethylene diisocyanate (HDI) induced a dose-dependent increase in serum IgE concentration relative to both concurrent vehicle-treated control groups and historical control values. No similar increases were observed when mice were treated in the same way with either DNCB or oxazolone (Hilton et al., 1995).

Dose selection for the mouse IgE test is an important issue. In practice the policy currently is to examine test chemicals first in the local lymph node assay, a mouse predictive test for the identification of contact allergens (Kimber and Basketter, 1992; Kimber et al., 1994). Wherever possible application concentrations are selected for use in the mouse IgE test that elicit a positive response in the local lymph node assay. The primary advantage of this strategy is in the interpretation of negative results in the mouse IgE test. If concentrations have been tested that yield a positive response in the local lymph node assay, then a negative result in the mouse IgE test can not be attributed to the failure of the chemical to gain access to the immune system when delivered via the skin, or to stimulate an active immune response. In such instances, as for example with DNCB, the interpretation is that the chemical is immunogenic and has the potential to induce contact sensitization, but fails to provoke the quality of immune response necessary for effective sensitization of the respiratory tract. A second potential advantage of performing a local lymph node assay in advance of the mouse IgE test is that the former may serve as an initial screen for potential chemical respiratory allergens. For reasons that are not entirely clear, all known human chemical respiratory sensitizers elicit positive responses in predictive tests for contact sensitization potential; this being despite the fact that many such chemicals only very rarely cause allergic contact dermatitis in humans (Kimber, 1995). As such it could be argued that chemicals that fail to stimulate a positive response in the local lymph node assay, or other predictive tests for skin sensitization, are unlikely to have the ability to induce respiratory allergy.

In the context of developing the mouse IgE test as a method for the routine predictive evaluation of respiratory sensitizing potential, consideration has been given recently to the need for positive and negative controls and the criteria for a positive response (Hilton et al., 1996). It is recommended presently that responses in the mouse IgE test are evaluated statistically using analysis of variance combined with the two-sided Student's t-test. A positive response is one in which the mean serum concentration is increased significantly in comparison with values derived from control mice treated concurrently with the relevant vehicle alone. On this basis, a chemical with the potential to induce sensitization of the respiratory tract is defined by the mouse IgE test as one which provokes, at one or more test concentrations, a significant elevation in the concentration of serum IgE (Hilton et al., 1996). The utility, in practice, of this criterion for a positive response will need to be evaluated carefully with a wider range of test chemicals. Nevertheless, at this time, such an approach appears appropriate and is favoured in preference to analysis of fold-increases in IgE concentrations or to comparisons with historical control values.

During the course of investigations of the mouse IgE test, responses induced by TMA and DNCB have been measured concurrently in over 15 independent experiments. The test concentrations chosen for such comparisons (25 per cent TMA and 1 per cent DNCB) were selected as those that were found to be of equivalent immunogenicity in BALB/c strain mice with respect to the stimulation of draining lymph node cell proliferative responses. In all experiments the concentration of serum IgE recorded following exposure of mice to TMA was increased significantly compared with levels measured in sera drawn from animals treated concurrently with vehicle alone or with DNCB. In no instance did DNCB cause an increase in serum IgE levels rela-

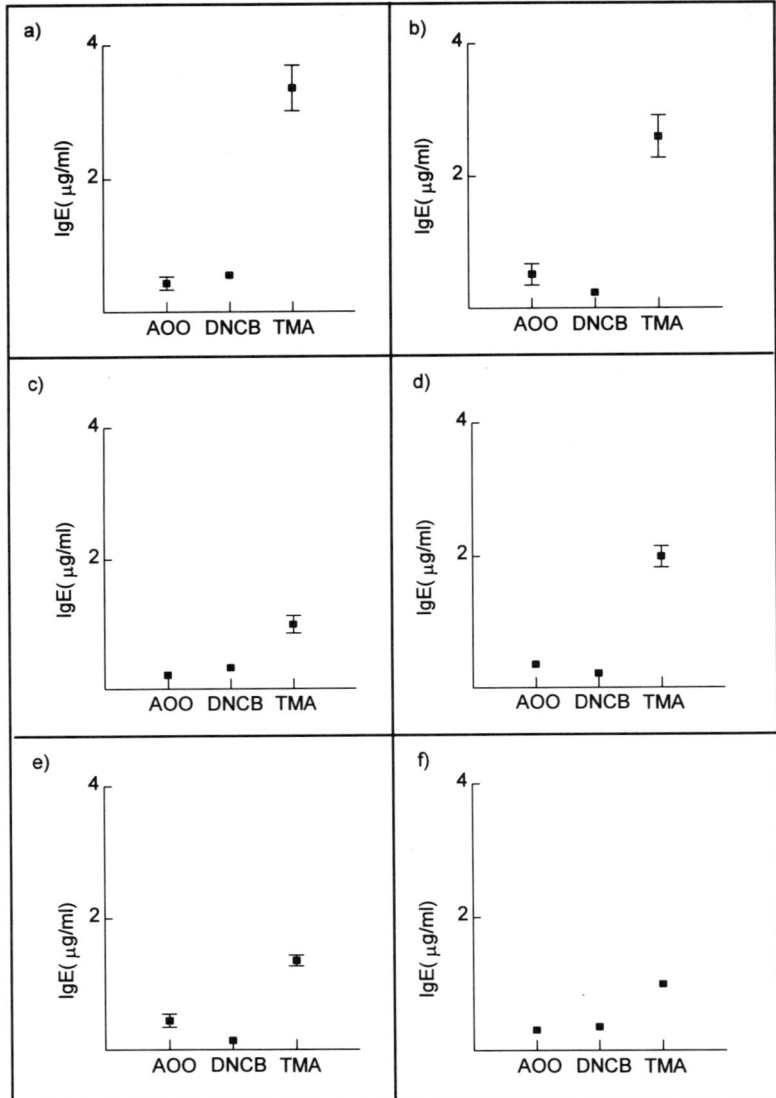

Figure 7.1 Analysis of serum IgE concentrations following exposure of mice to allergens or vehicle. Groups of mice ($n = 6$) were exposed to 50 µl of 25 per cent TMA or 1 per cent DNCB bilaterally on the shaved flanks. Control mice received identical treatment with vehicle (AOO) alone. Seven days later, 25 µl of the same test chemical at half the application concentration used previously, or an equal volume of vehicle alone, were applied to the dorsum of both ears. Fourteen days following the initiation of exposure, mice were exsanguinated by cardiac puncture and serum prepared. Serum IgE was measured using a sandwich ELISA. Results are expressed as mean serum IgE concentration in µg . ml^{-1} ± SE (where these exceeded 0.075 µg . ml^{-1}). A summary of six independent experiments a–f.

Assessment of Respiratory Sensitizing Potential of Chemicals in Mice

tive to vehicle controls (Hilton *et al.*, 1996). On the basis of these results it is recommended that TMA and DNCB at these concentrations should be incorporated in mouse IgE tests as, respectively, positive and negative controls. The results of six such independent experiments are summarized graphically in Figure 7.1. The data illustrated in Figure 7.1 reveal that there exists some inter-experimental variation, particularly with respect to the vigour of responses induced by TMA. It is also the case, but not apparent from the mean values illustrated in Figure 7.1, that variation in responses between mice within an individual experimental group is not uncommon. The reasons for such variations are not known, although it is likely that stress and sub-clinical infection may impact significantly on IgE production.

In conclusion, the experimental protocol recommended presently is as follows. Groups of six BALB/c strain mice are exposed topically on both shaved flanks to 50 µl of one of three concentrations of the test material or to an equal volume of vehicle alone. Additional groups of mice receive 25 per cent TMA, 1 per cent DNCB or the same volume of 4:1 acetone:olive oil, the vehicle of choice for these chemicals. Seven days later mice are treated on the dorsum of both ears with 25 µl of the same chemical at half the application concentration used previously, or with the same volume of vehicle alone. Fourteen days following the initiation of exposure, mice are exsanguinated by cardiac puncture and serum prepared and stored at −20°C until analysis (Hilton *et al.*, 1995, 1996). This protocol is illustrated diagrammatically in Figure 7.2.

To date there is little confirmatory evidence available from independent laboratories regarding the sensitivity and selectivity of the mouse IgE test, although Potter and Wederbrand (1995) have recently reported similar, although not identical, results using BALB/c strain mice. Of interest also are preliminary data that indicate a rat variant of the mouse IgE test may be achievable. It was shown that topical exposure to TMA, but not DNCB, elicited an increase in the serum concentration of IgE in Brown Norway rats (Arts *et al.*, 1995).

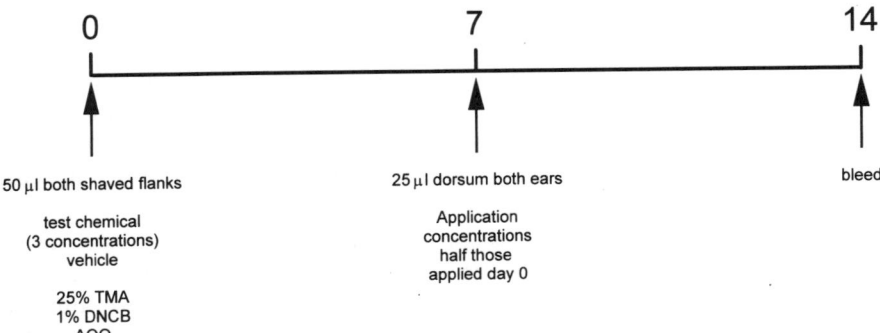

Figure 7.2 Schema illustrating current protocol for conduct of a mouse IgE test, incorporating 25 per cent TMA (positive control) and 1 per cent DNCB (negative control).

The advantages and disadvantages of the mouse IgE test as a method for identifying chemical respiratory allergens have been considered recently in different fora (Briatico-Vangosa *et al.*, 1994; Selgrade *et al.*, 1994). Advantages include the fact that, compared with guinea pig methods, the mouse IgE test is rapid, cost-effective and does not require the availability of sophisticated inhalation facilities or respiratory monitoring equipment. Moreover, there is no requirement in the mouse IgE test to elicit hypersensitivity reactions and as a consequence animals are subject to less trauma than those used in methods based upon elicitation of pulmonary responses. It might be possible also to use the mouse IgE test for consideration of potency as well as hazard identification. One suggestion is that estimation of the lowest concentration of the test material required to stimulate a significant increase in the serum concentration of IgE could be used as the basis for comparisons of relative potency and for derivation of a no-observable-effect level (NOEL).

Drawbacks and disadvantages of the mouse IgE test are as follows. The available data suggest that protein respiratory allergens do not induce in mice an increase in serum IgE concentration comparable with that associated with chemical allergens (Hilton *et al.*, 1994). In its present form at least, therefore, the mouse IgE test does not lend itself to assessment of the respiratory allergic potential of proteins. In addition, it is important to emphasize that the mouse IgE test seeks to identify those chemicals that exhibit the potential to induce immune responses of the quality required to cause sensitization of the respiratory tract. It is not necessarily the case, however, that the hazard identified in this way will translate into a significant risk of respiratory allergy in humans. It might be conjectured, for instance, that a chemical identified as positive in the mouse IgE test could lack the ability to gain access to, or to elicit hypersensitivity reactions in, the respiratory tract. An additional perceived constraint is that, unlike most guinea pig methods, the mouse IgE test can not be used to examine dose–response relationships for the elicitation of respiratory allergic reactions in previously sensitized animals; nor does the mouse IgE test provide a method for examination of cross-reactivity between chemical allergens. None of the above considerations serve to compromise the potential utility of the mouse IgE test as a screening method for the identification of chemical respiratory allergens. At an operational level, however, an important issue is the stability of IgE responses. As mentioned above, a number of factors influence the serum concentration of IgE and it is possible that these may serve, in certain circumstances, to compromise the performance of the assay. The current view is that successful conduct of the mouse IgE test will necessitate careful control of experimental conditions.

In summary, the mouse IgE test provides a novel and, to date, promising alternative method for the predictive identification of potential chemical respiratory allergens. The requirements now are for a more extensive evaluation of the assay with a wider range of test chemicals and for an examination of the method in independent laboratories.

7.4 Cytokine Fingerprinting

As described above, the divergent immune responses induced in mice by chemicals such as TMA and DNCB are characterized by selective cytokine secretion patterns by draining lymph node cells. Experience to date indicates that the phenotype of selective cytokine secretion evolves as the immune response to a chemical allergen matures. Draining lymph node cells isolated early during the immune response, following initial exposure of mice, produce similar levels of cytokines irrespective of the chemical allergen used. After more chronic treatment, however, selective cytokine secretion becomes apparent. Thus, for instance, after primary exposure of BALB/c mice to TMA or oxazolone, draining lymph node cells produce during culture similar amounts of mitogen-inducible IL-4. Following more prolonged exposure, however, the production of IL-4 is significantly greater by cells prepared from mice sensitized with TMA (Dearman et al., 1994, 1995). Exposure of mice to allergen over a 13-day period results also in substantial differences in the spontaneous production of IL-10 and IFN-γ; treatment with TMA favouring the production of the former, and oxazolone or DNCB the latter (Dearman et al., 1995). Results of a typical experiment in which cytokine responses induced following exposure of mice to TMA or DNCB are illustrated in Figure 7.3. The development with time of selective cytokine expression patterns is consistent with what is known of the development of functional subpopulations of Th cells. It is believed that Th1 and Th2 cells may be regarded as the most differentiated types of T-helper cells and that they develop during an immune response from a common precursor, designated Th0, that is able to elaborate both Th1- and Th2-type cytokines (Firestein et al., 1989, Street et al., 1990). It must be recognized, however, that not all the cytokines produced by allergen-activated lymph nodes necessarily derive from $CD4^+$ Th cells. There is increasing evidence that $CD8^+$ T-lymphocytes can also exhibit selective cytokine expression (Croft et al., 1994). Irrespective of the cellular source of the cytokines produced, it is clear that the patterns described above characterize immune responses provoked by other chemical allergens. Recent investigations have revealed that, under the same conditions of exposure, TDI and DNFB stimulate distinct cytokine patterns. Draining lymph node cells derived from mice treated with TDI were found to secrete high levels of IL-10 and mitogen-inducible IL-4, but only comparatively low concentrations of IFN-γ. Exposure to DNFB resulted in the converse profile of cytokine secretion (Dearman et al., 1996).

It is proposed therefore that while the stimulation of cytokine production *per se* is a reflection of the inherent immunogenicity of a chemical, the selectivity of cytokine secretion is a function of the quality of immune response provoked and, as a consequence, is indicative of the type of allergic reaction that will develop subsequently. If such is the case then it should be possible to classify chemical allergens on the basis of cytokine selectivity. Such 'cytokine fingerprinting' is particularly attractive as it would permit, in a single inte-

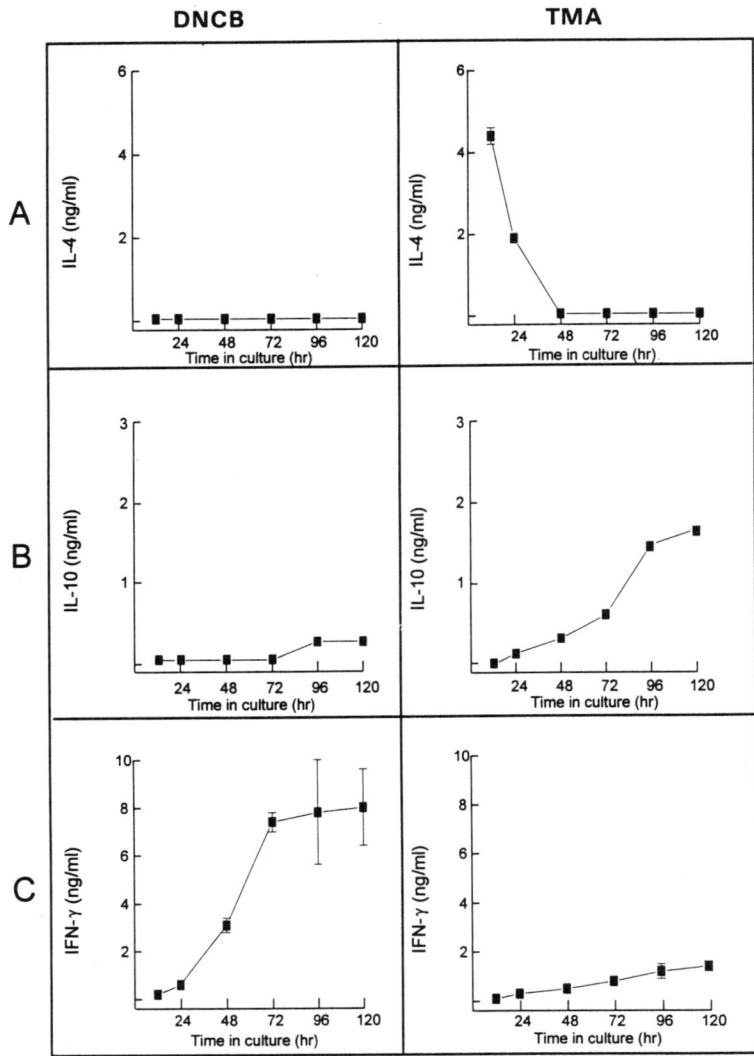

Figure 7.3 Production by draining lymph node cells of IL-4, IL-10 and IFN-γ following repeated topical exposure of mice to TMA or DNCB. Groups of BALB/c strain mice ($n = 10$) received 50 μl of 1 per cent DNCB or 10 per cent TMA (both in AOO) bilaterally on both shaved flanks. Five days later this treatment was repeated. After a further 5 days, 25 μl of chemical was applied to the dorsum of both ears daily for 3 consecutive days. One day following the final exposure mice were killed and draining auricular lymph nodes excised and pooled for each experimental group. A single cell suspension of lymph node cells was prepared and cultured in the presence (for mitogen-inducible IL-4 production), or absence (for the measurement of spontaneous IL-10 and IFN-γ secretion) of 2 μg . ml^{-1} concanavalin A. Culture was terminated after various periods and the concentrations of IL-4, IL-10 and IFN-γ measured in supernatants by cytokine-specific ELISA. In each case cytokine concentrations are recorded as mean values in ng . ml^{-1}. Standard errors are shown when greater than 0.3 ng . ml^{-1}. A, IL-4; B, IL-10 and C, IFN-γ.

grated assay, the simultaneous identification and characterization of chemical sensitizers.

There are real opportunities here and attention is focused currently on examination of cytokine secretion profiles following exposure of mice to a wider range of chemical allergens. Concurrently, investigations are proceeding to determine the extent to which differential cytokine secretion reflects transcriptional activation and whether analysis of the expression by lymph node cells of mRNA for selected cytokines would provide an alternative method of fingerprinting.

7.5 Conclusions

In this chapter we have described two novel approaches to the identification of chemical respiratory allergens in mice: the mouse IgE test and cytokine fingerprinting. Although both methods show promise, it would be fair to say that the development in mice of predictive tests for respiratory allergic potential is still in its infancy and that there is some way to go before a fully validated and widely accepted method is available. No animal model, whether in the mouse or other species, can be expected realistically to replicate accurately all aspects of human respiratory allergy and occupational asthma. It is nevertheless the case, however, that in practice the mouse represents a sound basis for the experimental dissection of the complex biology of immune responses to chemical allergens. The value of the mouse as a model of chemical allergy and the popularity of this species for basic immunological research should ensure that new opportunities for toxicological evaluation continue apace.

References

ARTS, J., KUPER, F., DROGE, S. & FERON, V. (1995) The local lymph node assay and the IgE test in the rat: an alternative animal model for airway hypersensitivity screening. Abstract presented at the Belgian Society for Toxicology, May 1995.

BERNSTEIN, D. I., PATTERSON, R. & ZEISS, C. R. (1982) Clinical and immunologic evaluation of trimellitic anhydride- and phthalic anhydride-exposed workers using a questionnaire with comparative analysis of enzyme-linked immunosorbent and radioimmunoassay studies. *Journal of Allergy and Clinical Immunology*, **69**, 311–318.

BOTHAM, P. A., RATTRAY, N. J., WOODCOCK, D. R., WALSH, S. T. & HEXT, P. M. (1989) The induction of respiratory allergy in guinea-pigs following intradermal injection of trimellitic anhydride: a comparison with the response to 2,4-dinitrochlorobenzene. *Toxicology Letters*, **47**, 25–39.

BRIATICO-VANGOSA, G., BRAUN, C. L. J., COOKMAN, G., HOFMANN, T., KIMBER, I., LOVELESS, S. E., MORROW, T., PAULUHN, J., SORENSEN, T. & NIELSEN, H. J. (1994) Respiratory allergy: hazard identification and risk assessment. *Fundamental and Applied Toxicology*, **23**, 145–158.

CHAN-YEUNG, M. (1995) Occupational asthma. *Environmental Health Perspectives*, **103** (Suppl. 6), 249–252.

CROFT, M., CARTER, L., SWAIN, S. L. & DUTTON, R. W. (1994) Generation of polarized antigen-specific CD8 effector populations: reciprocal action of interleukin 4 (IL-4) and IL-12 in promoting type 2 versus type 1 cytokine profiles. *Journal of Experimental Medicine*, **180**, 1715–1728.

DEARMAN, R. J. & KIMBER, I. (1991) Differential stimulation of immune function by respiratory and contact chemical allergens. *Immunology*, **72**, 563–570.

(1992) Divergent immune responses to respiratory and contact chemical allergens: antibody elicited by phthalic anhydride and oxazolone. *Clinical and Experimental Allergy*, **22**, 241–250.

DEARMAN, R. J., HEGARTY, J. M. & KIMBER, I. (1991) Inhalation exposure of mice to trimellitic anhydride induces both IgG and IgE anti-hapten antibody. *International Archives of Allergy and Applied Immunology*, **95**, 70–76.

DEARMAN, R. J., BASKETTER, D. A., COLEMAN, J. W. & KIMBER, I. (1992a) The cellular and molecular basis for divergent allergic responses to chemicals. *Chemical-Biological Interactions*, **84**, 1–10.

DEARMAN, R. J., BASKETTER, D. A. & KIMBER, I. (1992b) Variable effects of chemical allergens on serum IgE concentration in mice. Preliminary evaluation of a novel approach to the identification of respiratory sensitizers. *Journal of Applied Toxicology*, **12**, 317–323.

DEARMAN, R. J., MITCHELL, J. A., BASKETTER, D. A. & KIMBER, I. (1992c) Differential ability of occupational chemical contact and respiratory allergens to cause immediate and delayed dermal hypersensitivity reactions in mice. *International Archives of Allergy and Immunology*, **97**, 315–323.

DEARMAN, R. J., SPENCE, L. M. & KIMBER, I. (1992d) Characterization of murine immune responses to allergenic diisocyanates. *Toxicology and Applied Pharmacology*, **112**, 190–197.

DEARMAN, R. J., RAMDIN, L. S. P., BASKETTER, D. A. & KIMBER, I. (1994) Inducible interleukin-4-secreting cells provoked in mice during chemical sensitization. *Immunology*, **81**, 551–557.

DEARMAN, R. J., BASKETTER, D. A. & KIMBER, I. (1995) Differential cytokine production following chronic exposure of mice to chemical respiratory and contact allergens. *Immunology*, **86**, 545–550.

DEARMAN, R. J., BASKETTER, D. A. & KIMBER, I. (1996) Characterization of chemical allergens as a function of divergent cytokine secretion profiles in mice. *Toxicology and Applied Pharmacology*, **138**, 308–316.

DREXLER, H., SCHALLER, K.-H., WEBER, A., LETZEL, S. & LEHNERT, G. (1993) Skin prick tests with solutions of acid anhydrides in acetone. *International Archives of Allergy and Immunology*, **100**, 251–255.

FABBRI, L. M., MAESTRELLI, P., SAETTA, M. & MAPP, C. M. (1994) Mechanisms of occupational asthma. *Clinical and Experimental Allergy*, **24**, 628–635.

FIRESTEIN, G. S., ROEDER, W. D., LAXER, J. A., TOWNSEND, K. S., WEAVER, C. T., HOM, J. T., LINTON, J., TORBETT, B. E. & GLASEBROOK, A. L. (1989) A new murine CD4$^+$ T cell subset with an unrestricted cytokine profile. *Journal of Immunology*, **143**, 518–525.

GARSSEN, M., NIJKAMP, F. P., VAN DER VLIET, H. & VAN LOVEREN, H. (1991) T-cell-mediated induction of airway hypersensitivity in mice. *American Review of Respiratory Disease*, **144**, 931–938.

HILTON, J., DEARMAN, R. J., BASKETTER, D. A. & KIMBER, I. (1994) Serological responses induced in mice by immunogenic proteins and by protein respiratory allergens. *Toxicology Letters*, **73**, 43–53.

(1995) Identification of chemical respiratory allergens: dose–response relationships in the mouse IgE test. *Toxicology Methods*, **5**, 51–60.

HILTON, J., DEARMAN, R. J., BOYLETT, M. S., FIELDING, I., BASKETTER, D. A. & KIMBER, I. (1996) The mouse IgE test for identification of potential chemical respiratory allergens. Considerations of stability and controls. *Journal of Applied Toxicology*, **16**, 165–170.

HOLLIDAY, M. R., COLEMAN, J. W., DEARMAN, R. J. & KIMBER, I. (1993) Induction of mast cell sensitization by chemical allergens: a comparative study. *Journal of Applied Toxicology*, **13**, 137–142.

KAROL, M. H., TOLLERUD, D. J., CAMPBELL, T. P., FABBRI, L., MAESTRELLI, P., SAETTA, M. & MAPP, C. E. (1994) Predictive value of airways hyper-responsiveness and circulating IgE for identifying types of responses to toluene diisocyanate inhalation challenge. *American Journal of Respiratory and Critical Care Medicine*, **149**, 611–615.

KIMBER, I. (1995) Contact and respiratory sensitization by chemical allergens: uneasy relationships. *American Journal of Contact Dermatitis*, **6**, 34–39.

(1996) Mechanisms of chemical respiratory allergy, in VOS, J. G., YOUNES, M. & SMITH, E. (Eds). *Allergic Hypersensitivities Induced by Chemicals. Recommendations for Prevention*, pp. 59–75. Boca Raton: CRC Press.

KIMBER, I. & BASKETTER, D. A. (1992) The murine local lymph node assay: a commentary on collaborative studies and new directions. *Food and Chemical Toxicology*, **30**, 165–169.

KIMBER, I. & DEARMAN, R. J. (1992) The mechanisms and evaluation of chemically induced allergy. *Toxicology Letters*, **64/65**, 79–84.

(1993) Approaches to the identification and classification of chemical allergens in mice. *Journal of Pharmacological and Toxicological Methods*, **29**, 11–16.

(1994) Immune responses to contact and respiratory allergens, in DEAN, J. H., LUSTER, M. I., MUNSON, A. E. & KIMBER, I. (Eds). *Immunotoxicology and Immunopharmacology*, 2nd edn, pp. 663–679, New York: Raven Press.

KIMBER, I., DEARMAN, R. J., SCHOLES, E. W. & BASKETTER, D. A. (1994) The local lymph node assay: developments and applications. *Toxicology*, **93**, 13–31.

LAUERMA, A. I., AIOI, A. & MAIBACH, H. I. (1995) Topical *cis*-urocanic acid suppresses both induction and elicitation of contact hypersensitivity in BALB/c mice. *Acta Dermatologica et Venereologica*, **75**, 272–275.

MOLLER, D. R., GALLAGHER, J. S., BERNSTEIN, D. I., WILCOX, T. G., BURROUGHS, H. E. & BERNSTEIN, I. L. (1985) Detection of IgE-mediated respiratory hypersensitivity in workers exposed to hexahydrophthalic anhydride. *Journal of Allergy and Clinical Immunology*, **76**, 663–672.

MOSMANN, T. R. & COFFMAN, R. L. (1989) Heterogeneity of cytokine secretion patterns and functions of helper T cells. *Advances in Immunology*, **46**, 111–147.

MOSMANN, T. R., SCHUMACHER, J. H., STREET, N. F., BUDD, R., O'GARRA, A., FONG, T. A. T., BOND, M. W., MOORE, K. W. M., SHER, A. & FIORENTINO, D. F. (1991) Diversity of cytokine synthesis and function of mouse $CD4^+$ T cells. *Immunological Reviews*, **123**, 209–229.

NIELSEN, J., WELINDER, H., OTTOSSON, H., BENSRYD, I., VENGE, P. & SKERFVING, S. (1994) Nasal challenge shows pathogenetic relevance of specific IgE serum antibodies for nasal symptoms caused by hexahydrophthalic anhydride. *Clinical and Experimental Allergy*, **24**, 440–449.

POTTER, D. W. & WEDERBRAND, K. S. (1995) Total IgE antibody production in

BALB/c mice after dermal exposure to chemicals. *Fundamental and Applied Toxicology*, **26**, 127–135.

SALVAGGIO, J. E., BUTCHER, B. T. & O'NEIL, C. E. (1988) Occupational asthma due to chemical agents. *Journal of Allergy and Clinical Immunology*, **78**, 1053–1057.

SELGRADE, M. K., ZEISS, C. R., KAROL, M. H., SARLO, K., KIMBER, I., TEPPER, J. S. & HENRY, M. C. (1994) Workshop on status of test methods for assessing potential of chemicals to induce respiratory allergic reactions. *Inhalation Toxicology*, **6**, 303–319.

STREET, N. E., SCHUMACHER, J. H., FONG, T. A. T., BASS, H., FIORENTINO, D. F., LEVERAH, J. A. & MOSMANN, T. R. (1990) Heterogeneity of mouse helper T cells: evidence from bulk cultures and limiting dilution cloning for precursors of Th1 and Th2 cells. *Journal of Immunology*, **144**, 1629–1639.

ZEISS, C. R. & PATTERSON, R. (1993) Acid anhydrides, in BERNSTEIN, I. L., CHAN-YEUNG, M., MALO, J-L. & BERNSTEIN, D.I. (Eds). *Asthma in the Workplace*, pp 439–457. New York: Marcel Dekker.

8

Predictive Assessment of Respiratory Sensitizing Potential of Proteins in Mice

MICHAEL K. ROBINSON and **THOMAS T. KAWABATA**

The Procter & Gamble Company, Cincinnati, Ohio

8.1 Introduction

Various animal models have been developed over the years to aid the study of immediate hypersensitivity and respiratory responsiveness to protein allergens. These models have included both larger (e.g. horse, sheep, pig, monkey, dog) and smaller (e.g. rabbit, guinea pig, rat, mouse) animal species (Hirshman *et al.*, 1980; Abraham *et al.*, 1983; Derksen *et al.*, 1985; Mapp *et al.*, 1985; Ishida *et al.*, 1989; Elwood *et al.*, 1991; Renz *et al.*, 1992; Katz *et al.*, 1993; Minshall *et al.*, 1993; Mauser *et al.*, 1995), and have utilized naturally allergic animals (e.g. horse, sheep) (Abraham *et al.*, 1983; Derksen *et al.*, 1985), as well as both actively and passively sensitized animals (e.g. rabbits, guinea pigs, rats, mice) (Mapp *et al.*, 1985; Elwood *et al.*, 1991; Madi *et al.*, 1991; Renz *et al.*, 1992; Minshall *et al.*, 1993). For the most part, the literature on these model systems has focused on four areas: immunology/allergy (Sehon and Mohapatra, 1992; Renz *et al.*, 1994), functional effects (Elwood *et al.*, 1991; Misawa and Sugiyama, 1993), inflammatory pathophysiology (Elwood *et al.*, 1991; Tarayre *et al.*, 1992; Martin *et al.*, 1993; Coyle *et al.*, 1995), and therapeutics (Abraham, 1993; Wegner *et al.*, 1993). A major emphasis has been on the use of protein allergens to investigate mechanisms of (Elwood *et al.*, 1991) and therapeutic intervention in asthma and airway hyperreactivity (Wegner *et al.*, 1993).

For many years, the mouse has been a favoured animal to investigate the development and regulation of immunoallergic (IgE antibody) responses to many different protein allergens (Henderson *et al.*, 1985, 1987; Inagaki *et al.*, 1985; Takafuji *et al.*, 1987). Mouse strains are divided into high, moderate and low IgE responders based on the amount of specific IgE antibody produced after sensitization to various allergens (Revoltella and Ovary, 1969; Berzofsky *et al.*, 1977; Mancino and Ovary, 1980). The allergens used have included model proteins (e.g. ovalbumin) (McCaskill *et al.*, 1984), hapten–protein conjugates (Taylor et al., 1980), common respiratory allergens (e.g. rye grass) (McCaskill et al., 1982), and bacteria and parasites (or extracts derived from them) (Mitchell, 1976; Poirier et al., 1988). Different immunization regimens (routes, timing, frequency, adjuvants, etc.) have been used and assays have included cellular and humoral IgE responses, other antibody responses, cellular proliferation, antigen presentation and allergic inflammation.

More recently, the mouse has begun to gain favour as a model for respiratory allergy/asthma and airway responsiveness. Publications from several laboratories have documented increases in airflow resistance, airway hyperreactivity to cholinergic agonists and cellular inflammation in allergic mice challenged with the relevant allergenic proteins (or chemicals) via the airways (Garssen et al., 1991; Renz et al., 1992; Lack et al., 1994; Kung et al., 1995). This has facilitated and will continue to aid the investigation of immunoinflammatory mechanisms of respiratory allergy and asthma because of the availability of reagents (e.g. cytokines, monoclonal antibodies) and procedures (e.g. transgenics) to modulate endogenous responses or processes (Brusselle et al., 1995; Kung et al., 1995).

Our interest in a mouse model of respiratory allergy to proteins was driven by the desire to develop a simple method to assess the relative potency of enzymes as respiratory allergens. The application of threshold limit values (TLV) and operational exposure guidelines for detergent enzymes has minimized sensitization and allergic symptoms are a rare event (Gaines, 1994). The TLV of the most widely used detergent enzyme, Alcalase® (protease subtilisin Carlsberg), was set empirically in the manufacturing plants (American Conference of Governmental Industrial Hygienists, 1990). To determine the operational guidelines of new enzymes that may be more or less allergenic than Alcalase, a guinea pig intratracheal test (GPIT) was developed (Ritz et al., 1993) (see also Chapter 6). The GPIT test has been a very useful method for the evaluation of enzyme allergenic potential (Sarlo et al., 1991); however, it is also very time consuming and expensive. Recently, Hilton et al. (1994) have demonstrated enzyme sensitization in mice after parenteral intraperitoneal (i.p.) exposure. However, we wanted to evaluate exposure regimes more relevant to those occurring in occupational settings. Therefore, our work in the mouse has been focused on developing quicker and less expensive methods employing sensitization via the respiratory tract.

This chapter summarizes our recent efforts to establish two different protein/enzyme dosing models in mice, using intratracheal and intranasal dosing. Both models have been used to measure antigen (enzyme) specific antibody responses as a function of the dose of antigen administered, the immunization procedure and regimen, and the isotype of antibody produced.

8.2 Intratracheal Model

The intratracheal (i.t.) model was developed for direct comparison to the GPIT method (Kawabata et al., 1996). The initial objective was to assess IgE and IgG1 antibody responses in the serum of mice administered a protease enzyme by the i.t. route. Since detergent plant workers are exposed to airborne enzymes in a detergent matrix, we also examined the effects of detergent matrix in enhancing the specific IgE and IgG1 responses to enzyme after i.t. exposure. The results of these studies demonstrated that the i.t. route of administration is an effective method in producing enzyme specific IgE and IgG1 responses in mice and that these responses are enhanced by co-exposure to detergent matrix. This latter finding is consistent with prior observations in guinea pigs (Markham and Wilkie, 1976, 1979) and monkeys (Cashner et al., 1980).

Mice were i.t. dosed by using a modification of the methods reported by Starcher and Williams (1989) and Sikorski et al. (1989). Anaesthetized BALB/c mice were placed on a Plexiglas restrainer (30° angle). Using a fibre optic light (Cole Palmer, Chicago, IL) taped to a tongue depressor to illuminate the epiglottis, a 23 gauge (blunt) angled needle attached to a 500 μl Hamilton glass syringe was inserted past the epiglottis and into the trachea. Once in the trachea, the 500 μl syringe was replaced with a 250 μl syringe filled with the dosing solution (100 μl). The solution was injected into the trachea. The syringe was removed and 100 μl of air was subsequently injected to push the remaining dosing solution into the lung. Mice were i.t. dosed once a week for 2 to 8 consecutive weeks. Five days after the last dose, mice were anaesthetized, exsanguinated, and serum was collected for IgE and IgG1 assays.

Alcalase-specific IgG1 in mouse serum was measured by using an antigen capture enzyme-linked immunosorbent assay (ELISA) (Kawabata et al., 1996). Alcalase coated immunoassay plates were washed and incubated with serially diluted serum samples, then developed with alkaline phosphatase conjugated goat anti-mouse IgG1 antibody and p-nitrophenyl phosphate substrate. The plates were read at 405/650 nm and the optical density (O.D.) values transformed by linear regression using Softmax™ software (Molecular Devices). Titres of each serum sample were calculated as the reciprocal dilution producing an O.D. reading of 0.5 (interpolated from regression line) and were expressed as a log 2 value.

Alcalase-specific IgE in mouse serum was measured by the rat basophil leukaemia (RBL-2H3) cell serotonin release assay initially developed for measuring anti-ovalbumin responses as described by Kawabata and Babcock (1993). Serotonin prelabelled RBL-2H3 cells were incubated with serial dilutions of serum samples followed by incubation with Alcalase protein. Released serotonin was measured and percentage release determined relative to positive and negative control samples. Specific IgE titre was defined as the highest dilution of test sample with a percentage release value greater than or equal to twice the percentage release produced by cells exposed to normal serum and antigen. Titres of each serum sample were calculated as a reciprocal dilution and expressed as a log 2 value. This method, optimized for measuring anti-Alcalase IgE titres, showed a strong correlation with *in vivo* passive cutaneous anaphylaxis responses (Kawabata et al., 1996).

Toxicology of Chemical Respiratory Hypersensitivity

A summary of results from several i.t. dosing studies with Alcalase is presented in Figure 8.1. Several variables had to be investigated in sequential studies in order to optimize the immunogenic (IgG1) and allergenic (IgE) responses to i.t. Alcalase exposure. These included the dose of Alcalase administered, the presence or absence of detergent in the dosing solution, and the number of doses given, among others. The detergent matrix used in these studies was a standard formula containing anionic and non-ionic surfactants, silicate builders and perborate bleach. Figure 8.1A shows the results of an early study in which two doses of Alcalase were administered in saline or admixed with either of two different amounts of detergent. The dosing material was given four times with 1 week separating each dose, a regimen that begins to produce significant antibody responses (after 4 weeks of dosing) in the GPIT (Ritz et al., 1993).

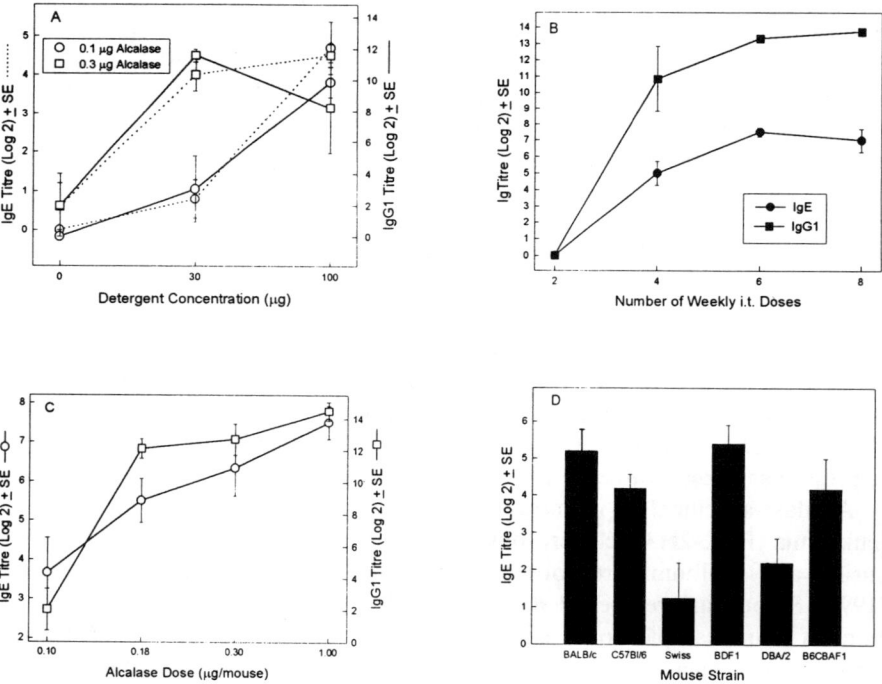

Figure 8.1 Mice received between four and eight weekly i.t. doses of 100 µl of Alcalase in saline or detergent matrix. The detergent matrix consisted of anionic and non-ionic surfactants, silicate builders and perborate bleach. Blood samples were obtained from individual mice 5 days after the final i.t. dose. The blood samples were assayed for anti-Alcalase specific IgG1 and IgE, respectively, by an ELISA method and by the RBL 2H3 cell ^{3}H-serotonin release assay as previously described (Kawabata et al., 1996). The mice used for Figures A–C were BDF1 strain. (Data reprinted with permission from Kawabata et al., 1996.)

For either dose of Alcalase tested (0.1 and 0.3 µg per mouse), both IgE and IgG1 titres were negligible in the absence of detergent. There was a dramatic adjuvant effect of detergent on both the IgE and IgG1 responses. This was seen at 30 µg of detergent when the higher (0.3 µg) Alcalase dose was administered and at both Alcalase doses in the presence of the 100 µg dose of detergent. In a separate experiment (not shown) an Alcalase dose of 0.6 µg in saline produced an IgE titre (4.0 ± 0.55) that was similar to that seen at lower Alcalase doses in detergent. Thus, an anti-Alcalase IgE response could be produced in the absence of detergent, but there was a two- to sixfold increase in potency when administered in the presence of 30 or 100 µg of detergent.

In the GPIT, weekly dosing of enzyme is carried out for 10 weeks with the response generally reaching a plateau at 6 to 8 weeks depending on the Alcalase dose administered (Ritz et al., 1993). In the mouse model, the IgE response plateaued at six weekly doses (Figure 8.1B) when Alcalase was administered at a dose of 0.3 µg/mouse in the presence of 100 µg/mouse of detergent. A similar pattern was seen with the IgG1 response, although the anti-Alcalase IgG1 titres were much higher. Using the 4-week dosing regimen, we compared the Alcalase dose response for both the IgE and IgG1 antibody isotypes when Alcalase was administered with 100 µg of detergent. Figure 8.1C shows a linear increase in the IgE titre with increasing Alcalase dose. The IgG1 response showed a steeper dose response curve almost reaching a plateau at a dose of 0.18 µg/mouse.

There was some variability seen in the ability of different mouse strains to produce an anti-Alcalase IgE response. When 0.3 µg Alcalase was administered (4 weekly doses) with 100 µg detergent, BALB/c, C57BL/6, BDF1 and B6CBAF1 mice all showed statistically equivalent responses (Figure 8.1D). The Swiss and DBA/2 mice showed significantly lower responses than the other strains. This difference in responsiveness was not solely H-2 dependent since BALB/c and DBA/2 mice share the H-2^d haplotype. Strain variations in responsiveness have been seen with other protein allergens such as ovalbumin/ovomucoid (Holt et al., 1981), indicating that enzymes are not unique in this respect. Variables such as type of protein, adjuvant used, isotype of antibody and route of administration will all impact on strain susceptibility (Revoltella and Ovary, 1969; Petty and Steward, 1977; Kudo et al., 1978; Mancino and Ovary, 1980; Holt et al., 1981; Staruch and Wood, 1982; Nakashima et al., 1983; Gurvich and Korukova, 1986; Else and Wakelin, 1989; Tomlinson et al., 1989). This is further manifest in a different pattern of strain susceptibility observed with the intranasal model (below).

The results of these studies demonstrate that the i.t. route of exposure is a feasible method to examine the IgE and IgG1 responses to enzymes in mice. The specific IgG1 response to Alcalase followed the IgE dose–response relationship, but was much easier to measure (ELISA compared with RBL). It is not known if the antibody responses generated with the i.t. route would be similar to that produced with inhalation exposure at comparable doses. However, it is known that similar antibody responses to Alcalase via each exposure route were observed in the guinea pig (Ritz et al., 1993), which would presumably extrapolate to the mouse.

Some initial studies were conducted using the i.t. method to compare IgE responses to different enzymes. The results of a typical study will be presented below in conjunction with the intranasal model. While gross differences in potencies were detectable, it was difficult to quantify these differences because of the somewhat narrow window between minimal and maximal IgE responses discernible by the RBL assay and some resulting variability in Alcalase dose–response patterns between studies. It was decided, therefore, to use the IgG1 ELISA assay as the method of choice for determining the antibody response to enzyme exposure. Since IgE (not IgG1) is the primary allergic antibody isotype in mice (Revoltella and Ovary, 1969; Mancino and Ovary, 1980), the use of IgG1 as a surrogate assay endpoint would make this (strictly speaking) an immunogenicity rather than an allergenicity model. Due to the additional factor of technical difficulty of the dosing procedure, we further shifted attention to development of an intranasal dosing model as a potentially simpler method to evaluate comparative antibody responses to enzymes.

8.3 Intranasal Model

The intranasal (i.n.) route has often been used to immunize mice to a variety of viral (Muller and Brons, 1994; Walsh, 1994), bacterial (Garcia et al., 1993, 1994), fungal (Wang et al., 1994) and soluble protein (Henderson et al., 1985; Takafuji et al., 1989) antigens. The methodology also offered a much simpler (and less costly) alternative to the i.t. dosing regimen.

Others have demonstrated that i.n. administration of proteins to mice (McCaskill et al., 1982; Henderson et al., 1985; Inagaki et al., 1985; Takafuji et al., 1989) and rats (Hameleers et al., 1991) can elicit specific IgE and/or IgG responses. However, studies were needed to determine the feasibility of using an i.n. route of administration in mice as an alternative test for measuring systemic antibody responses to enzymes. Various studies were conducted to examine different mouse strains and dosing regimens, as well as an investigation of the kinetics of the response, the isotype of antibody produced and the effects of detergent matrix. The results of these studies demonstrate that the i.n. dosing method is a highly practical and robust alternative method for assessing enzyme immunogenicity (Robinson et al., 1996). It represents a potential alternative to guinea pig methods and is also much easier, less time-consuming, and thereby less expensive to conduct, than the mouse i.t. method.

The intranasal dosing procedure used for these studies was very simple. Mice were anaesthetized and held in the palm of the hand, backs down. They were dosed intranasally (held upright or in a slight downward position) with various concentrations of Alcalase (0.01 to 20 µg protein), in saline or in a detergent matrix. Dosing solutions were administered in volumes of 10 µl per mouse and divided equally between the two nostrils. The dosing solutions were gently placed on the outside of each nostril and inhaled by the mouse. The mice were dosed in this manner on days 1, 3 and 10. Collection of sera was performed 5 days after the last instillation of antigen. This is the standard dosing regimen, although, in some studies, we have looked at alternative regimens. Sera are analyzed for enzyme specific IgG1 and IgE titres by the procedures described above for the i.t. method.

Initial studies in BALB/c mice failed to demonstrate a consistent antibody response (IgE or IgG1) after intranasal dosing by the above regimen. While it may have been possible to induce a better response in these mice by prolonging the dosing regimen, it was decided instead to examine other strains of mice to see if any would respond using the day 1/3/10 dosing regimen. BDF1 mice (a hybrid strain of C57Bl/6 and DBA/2 parents) showed a robust anti-Alcalase IgG1 dose response that titred out over a dose range of approximately 1.0 to 0.1 µg/mouse (Figure 8.2A). The antibody titre closely paralleled the percentage of responding animals at each dose administered. In these early studies, the enzyme was administered in detergent (30 µg per mouse), because of the adjuvant effects that had been seen in the i.t. dosing studies.

As in the i.t. method, the effect of i.n. administration of enzyme in a detergent matrix was to shift the dose–response curve to the left. This effect is shown in Figure 8.2B which summarizes multiple studies both with and without detergent administration. The potency of Alcalase in producing an anti-Alcalase IgG1 response was approximately fourfold greater in the presence of detergent as determined by comparing the ED_{50} values. The ED_{50} value (i.e. dose of enzyme producing an antibody response halfway between the maximum and minimum responses obtained for each dose response data set) was generated by the curve fitting process. Examination of several different strains of mice revealed BDF1 and CB6F1 (BALB/c crossed with C57Bl/6) mice were high responders whereas BALB/c mice and outbred CD1 mice were low responders (Figure 8.2C). Given the common $H-2^d$ haplotype of the BALB/c and DBA/2 strains, it was possible that high responsiveness was correlated with the $H-2^b$ haplotype of the C57Bl/6 parent. When C57Bl/6 mice were tested, they too were high responders. This pattern was different from what was seen in the i.t. model, where BALB/c mice were high IgE responders. As noted above, this might be attributable to the different route of administration or the isotype of antibody assayed.

The intranasal model, as currently developed, should be considered an immunogenic (rather than allergenic) model because of the fact that the antibody analyzed (IgG1) is not known to be involved in triggering allergic manifestations in mice. As indicated above, IgE is the major allergenic antibody in mice. Under the standard test conditions used for intranasal studies, mice fail to produce a strong or consistent IgE response. Extending the dosing procedure to 2 months (nine weekly doses) an IgE response becomes detectable (Figure 8.2D). As also seen in Figure 8.2D, administering two additional doses of enzyme further increased the maximal IgG1 titre produced, which then remained at the same high level with further dosing.

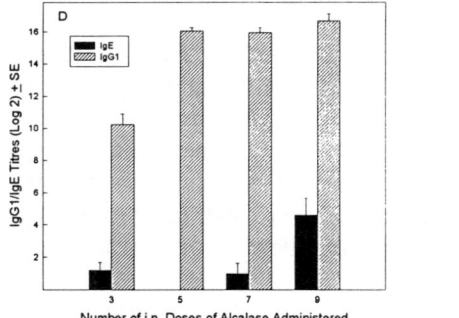
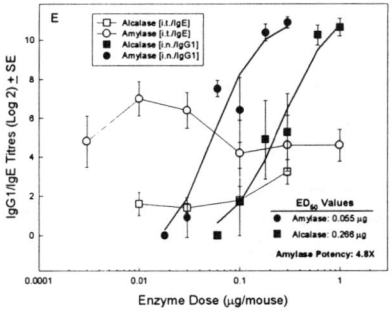

Figure 8.2 Mice received doses of Alcalase or amylase by the i.n. route (Figures A–E) or the i.t. route (Figure E). For the i.n. dosing, the mice received 5 μl of enzyme solution (in saline or detergent matrix) in each nostril on days 1, 3, and 10. For Figure 2D, additional weekly doses were administered. The mice used for Figures 2A, B, D, and E were all BDF1 strain. The i.t. enzyme shown in Figure 2E was administered for four weekly doses as indicated in the legend to Figure 8.1. Blood was obtained from individual mice 5 days after the final i.n. or i.t. dose. The blood samples were assayed for anti-enzyme specific IgG1 and IgE, respectively, by an ELISA method and by the RBL 2H3 cell ^{3}H-serotonin release assay as previously described (Kawabata et al., 1996). (Data reprinted with permission from Robinson et al., 1996.)

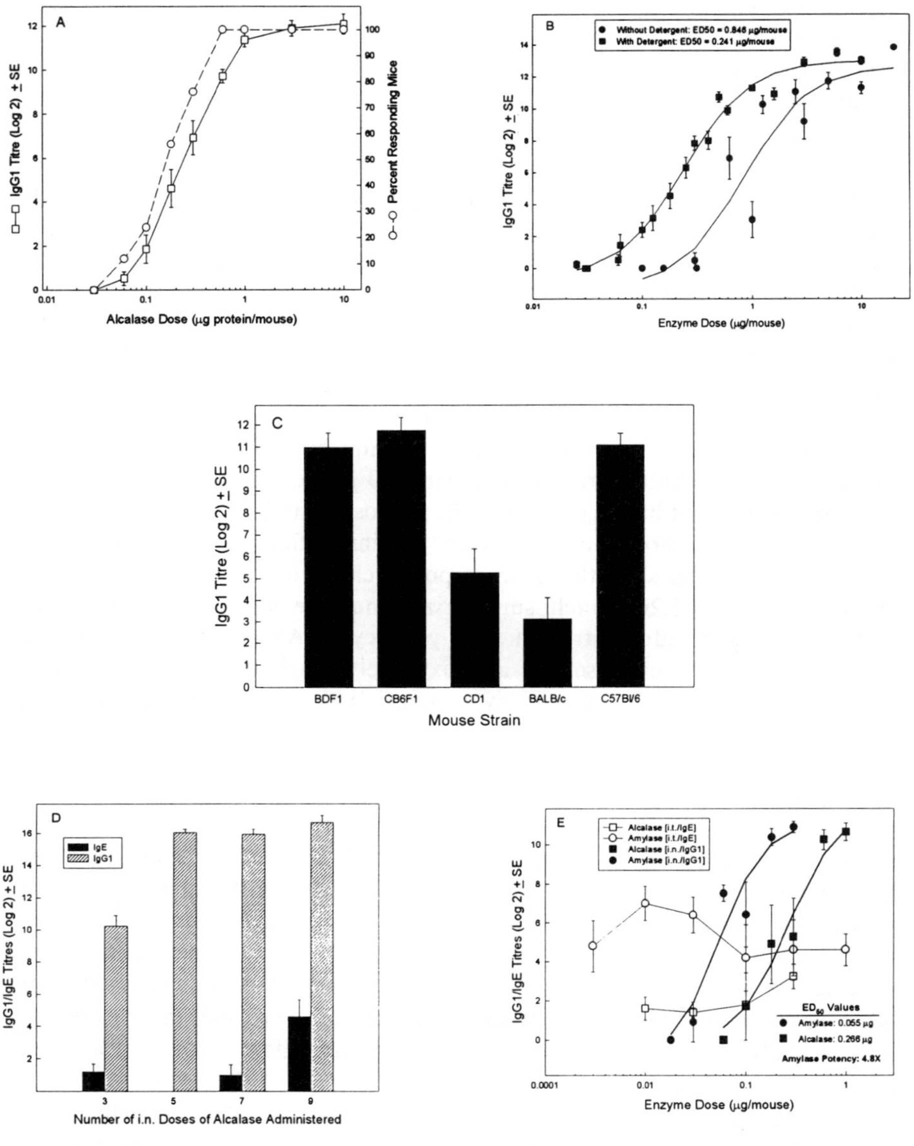

Assessment of the value of the intranasal model as a means of comparing the immunogenicities of different enzymes (or other allergenic proteins) is at a very early stage. Data from the GPIT test indicate that different enzymes have different potencies when compared with Alcalase (Sarlo *et al.*, 1991) and it will be important to show that the i.n. model can also detect these differences. Several enzymes of different classes have been tested in the intranasal model and have also shown a range of potencies (lower, equal, greater) compared with Alcalase using the relative ED_{50} values as the indicator of magnitude of the antibody response. In each case, the enzymes have been tested over a range of doses and the sera analyzed by protein-specific ELISA assays optimized for each enzyme. The range of antibody titres are comparable for all of the enzymes (i.e. log 2 is approximately 0–13) under the day 1/3/10 dosing regimen. The less potent ones simply require larger doses to develop full antibody responses. An example of a potency comparison study (Alcalase compared with amylase) for both the intratracheal and intranasal methods is shown in Figure 8.2E.

In this study, mice received Alcalase or a test amylase at several doses in detergent matrix by either the i.n. route or the i.t. route. For the i.t. study, the mice were given four weekly doses and blood was collected 5 days after the last dose. Enzyme-specific IgE was measured in each serum sample by the RBL 2H3 cell ^{3}H-serotonin release assay. It is difficult to quantify precisely the difference in potency, because of the flat dose–response curves seen for both Alcalase and amylase in the i.t./IgE study. This difficulty is further compounded by the variability seen in the titres of anti-Alcalase IgE seen from study to study (e.g. compare Alcalase titres in Figure 8.1C and 8.2E over the same doses administered). This variability may be due to differences in the biological responses in the animals and/or differences in the sensitivity of the RBL assay from study to study.

In the i.n. study, mice were given Alcalase or amylase in detergent on days 1, 3, and 10 and blood was collected on day 15. Here, enzyme-specific IgG1 was assayed by an ELISA. In the i.n. study, a set of parallel dose–response curves was obtained that covered the range of possible titres from no response to a plateau response for each enzyme. When logistics curves were fitted to these two data sets, a potency difference of 4.8 was determined by comparison of ED_{50} values. An encouraging note is that prior GPIT testing of this enzyme established a three- to tenfold greater potency than Alcalase in that model (Sarlo *et al.*, 1991).

While the use of IgG1 as an immunogenicity endpoint to attempt to predict the relative allergenicity of enzymes/proteins is open to criticism as mechanistically irrelevant, the true value of the method will lie in its performance relative to other preclinical test methods and human sensitization patterns. Many newly emerging methods in immunotoxicology rely on surrogate endpoints or endpoints that measure only a component of the response process (e.g. skin equivalent cultures for skin and eye irritation, local lymph node assay for contact sensitization, mouse IgE test for chemical respiratory sensitization).

Toxicology of Chemical Respiratory Hypersensitivity

Their value lies in their predictive capability, not in their precise pathophysiological or mechanistic adherence to the clinical disease or response process. Preliminary studies to date comparing Alcalase potency to the potency of several classes of enzymes (cellulase, amylase, lipase, protease) have shown similarity to prior GPIT test results (Horn *et al.*, 1996). Reports from others have shown strong linkage between the IgG1 and IgE-responses to protein allergens (Beck and Spiegelberg, 1989) providing additional support for its use as a surrogate marker. Also, there is recent evidence for IgE-independent hypersensitivity responses in mice (Oettgen *et al.*, 1994; Lei *et al.*, 1996), perhaps associated with an IgG1 isotype-mediated mechanism (Lei *et al.*, 1996; Oshiba *et al.*, 1996). Therefore, as long as the mouse i.n. model and IgG1 assay are capable of accurately determining relative enzyme immunogenic potency, the use of the IgG1 assay as a surrogate endpoint will be justified.

In addition to its simplicity in determining a precise differential potency between enzymes, the consistency of the dose–response pattern for Alcalase in the mouse i.n. model has been high from study to study. This is apparent in the narrow range of responses seen in Figures 8.2A and 8.2B (which show data compiled from multiple studies) and also in a very narrow range of ED_{50} values obtained from study to study (less than 20 per cent variability in ED_{50} values obtained from study to study). It is critical for any predictive methodology that the benchmark materials produce consistent responses. This is clearly the case for Alcalase in the i.n. model and supports the use of this enzyme and this test method in future studies.

8.4 Concluding Comments

The development of the intratracheal and intranasal models was driven by the objective to provide simpler and cost-effective respiratory hypersensitivity models in mice as alternative methods to evaluate the respiratory sensitization potential of enzymes. The intratracheal model offered advantages to the GPIT in overall reduced cost and time. It mimicked the GPIT in terms of animal exposure and used allergic IgE antibody production as the major isotype for evaluating enzyme potency. The major limitations of the intratracheal model rest with its relative technical difficulty and the variability around the IgE assay. Dosing mice by the i.t. route is technically demanding, time-consuming and therefore adds to the cost of the method. The IgE assay used is also subject to significant interstudy variability. As a functional cell-based assay, it is subject to normal types of variability inherent to cellular assays (cell viability, cell leakiness, isotope uptake, Ig receptor expression, etc.) which may vary due to subtle differences in incubation and culture conditions. As a result of this, the i.t. dosing/IgE assay procedure has not provided the study-to-study consistency and precise determination of enzyme potency differences needed to be considered a viable alternative to the GPIT test. Use of IgG1 as a surrogate endpoint would probably have improved our ability to distinguish (immunogenic) potencies, however, the i.t. method still had the disadvantage of a longer duration and more difficult dosing procedure.

The mouse intranasal model, in contrast, offers significant practical advantages over the GPIT and mouse i.t. methods in terms of simplicity, time commitment and cost. An additional advantage of the method lies in the fact that the exposure route mimics the most common human exposure. The method is simple to conduct and is relatively fast (2 to 3 weeks per study compared with 7 weeks for the mouse i.t., and 3 to 4 months for the GPIT). The IgG1 data obtained from repeated studies are consistent and potency differences between enzymes are easily quantifiable. In addition to the value of the method in routine safety evaluation, the faster and cheaper methodology could give enzyme formulators an opportunity to screen multiple candidate enzymes and factor immunogenic potency earlier into the development process.

Extensive effort is ongoing to develop further and characterize the intranasal model. This includes continued mechanistic investigations on the IgE response potential of the model, characterization of respiratory and inflammatory pathophysiology, responsiveness to enzymes delivered in different product matrices, and responses to individual enzymes when delivered in combination with other enzymes. As these results are forthcoming, they will be compared with prior results obtained from GPIT studies and with human experience in our efforts to validate the model fully and establish it as an appropriate and useful procedure for prediction of the respiratory hypersensitivity potential of new enzymes.

References

ABRAHAM, W. M. (1993) Effect of nedocromil sodium on specific mediator- and antigen-induced airway responses in allergic sheep. *Journal of Allergy and Clinical Immunology*, **92**, 171–176.

ABRAHAM, W. M., DELEHUNT, J. C., YERGER, L. & MARCHETTE, B. (1983) Characterization of a late phase pulmonary response after antigen challenge in allergic sheep. *American Review of Respiratory Disease*, **128**, 839–844.

AMERICAN CONFERENCE OF GOVERNMENTAL INDUSTRIAL HYGIENISTS (1990) *Threshold limit values and biological exposure indices*, p. 38. Cincinnati, OH.

BECK, L. & SPIEGELBERG, H. L. (1989) The polyclonal and antigen-specific IgE and IgG subclass response of mice injected with ovalbumin in alum or complete Freund's adjuvant. *Cellular Immunology*, **123**, 1–8.

BERZOFSKY, J. A., SCHECHTER, A. N., SHEARER, G. M. & SACHS, D. H. (1977) Genetic control of the immune response to staphylococcal nuclease. III. Time-course and correlation between the response to native nuclease and the response to its polypeptide fragments. *Journal of Experimental Medicine*, **145**, 111–122.

BRUSSELLE, G., KIPS, J., JOOS, G., BLUETHMANN, H. & PAUWELS, R. (1995) Allergen-induced airway inflammation and bronchial responsiveness in wild-type and interleukin-4-deficient mice. *American Journal of Respiratory Cell and Molecular Biology*, **12**, 254–259.

CASHNER, F., SCHUYLER, M., FLETCHER, R., RITZ, H. & SALVAGGIO, J. (1980) Immunologic responses of cynomolgus monkeys after repeated inhalation exposures to enzymes and enzyme-detergent mixtures. *Toxicology and Applied Pharmacology*, **52**, 62–68.

COYLE, A. J., ERARD, F., BERTRAND, C., WALTI, S., PIRCHER, H. & LE GROS, G. (1995) Virus-specific $CD8^+$ cells can switch to interleukin 5 production and induce airway eosinophilia. *Journal of Experimental Medicine*, **181**, 1229–1233.

DERKSEN, F. J., ROBINSON, N. E., ARMSTRONG, P. J., STICK, J. A. & SLOCOMBE, R .F. (1985) Airway reactivity in ponies with recurrent airway obstruction (heaves). *Journal of Applied Physiology*, **58**, 598–604.

ELSE, K. & WAKELIN, D. (1989) Genetic variation in the humoral immune responses of mice to the nematode *Trichuris muris*. *Parasite Immunology*, **11**, 77–90.

ELWOOD, W., LOTVALL, J. O., BARNES, P. J. & CHUNG, K. F. (1991) Characterization of allergen-induced bronchial hyperresponsiveness and airway inflammation in actively sensitized Brown Norway rats. *Journal of Allergy and Clinical Immunology*, **88**, 951–960.

GAINES, W. G. (1994) Occupational health experience manufacturing multiple enzyme detergents and methods to control enzyme exposure. *Proceedings of Toxicology Forum*, 143–147.

GARCIA, V. E., GHERARDI, M. M., IGLESIAS, M. F., CERQUETTI, M. C. & SORDELLI, D. O. (1993) Local and systemic immunity against *Streptococcus pneumoniae*: humoral responses against a non-capsulated temperature-sensitive mutant. *FEMS Microbiology Letters*, **108**, 163–167.

GARCIA, V. E., IGLESIAS, M. F., CERQUETTI, M. C., GOMEZ, M. I. & SORDELLI, D. O. (1994) Interaction between granulocytes and antibodies in the enhancement of lung defenses against *Staphylococcus aureus* after intranasal immunization of mice with live-attenuated bacteria. *FEMS Immunology and Medical Microbiology*, **9**, 55–63.

GARSSEN, J., NIJKAMP, F. P., VAN DER VLIET, H. & VAN LOVEREN, H. (1991) T-cell-mediated induction of airway hyperreactivity in mice. *American Review of Respiratory Disease*, **144**, 931–938.

GURVICH, A. E. & KORUKOVA, A. (1986) Induction of abundant antibody formation with a protein–cellulose complex in mice. *Journal of Immunological Methods*, **87**, 161–167.

HAMELEERS, D. M., VAN DER VEN, I., BIEWENGA, J. & SMINIA, T. (1991) Mucosal and systemic antibody formation in the rat after intranasal administration of three different antigens. *Immunology and Cell Biology*, **69**, 119–125.

HENDERSON, D. C., MORAN, D. M. & WHEELER, A. W. (1985) Differential suppressive influence of intranasal application of rye grass pollen extract on IgE antibody production in the mouse. *Clinical and Experimental Immunology*, **59**, 343–350.

HENDERSON, D. C., WHEELER, A. W. & MORAN, D. M. (1987) Suppression of murine IgE responses with amino acid polymer/allergen conjugates. V. By intranasal administration. *International Archives of Allergy and Applied Immunology*, **82**, 208–211.

HILTON, J., DEARMAN, R. J., BASKETTER, D. A. & KIMBER, I. (1994) Serological responses induced in mice by immunogenic proteins and by protein respiratory allergens. *Toxicology Letters*, **73**, 43–53.

HIRSHMAN, C. A., MALLEY, A. & DOWNES, H. (1980) Basenji–Greyhound dog model of asthma: reactivity to *Ascaris suum*, citric acid and methacholine. *Journal of Applied Physiology*, **49**, 953–957.

HOLT, P. G., ROSE, A. H., BATTY, J. E. & TURNER, K. J. (1981) Induction of adjuvant-independent IgE responses in inbred mice: primary, secondary and persistent IgE responses to ovalbumin and ovomucoid. *International Archives of Allergy and Applied Immunology*, **65**, 42–50.

HORN, P. A., BABCOCK, L. S., KAWABATA, T. T. & ROBINSON, M. K. (1996) Further development and characterization of a mouse intranasal method for comparative immunogenicity testing of enzymes. *The Toxicologist*, **30**, 95.

INAGAKI, N., TSURUOKA, N., GOTO, S., MATSUYAMA, T., DAIKOKU, M., NAGAI, H. & KODA, A. (1985) Immunoglobulin E antibody production against house dust mite, *Dermatophagoides farinae*, in mice. *Journal of Pharmacobiodynamics*, **8**, 958–963.

ISHIDA, K., KELLY, L. J., THOMSON, R. J., BEATTIE, L. L. & SCHELLENBERG, R. R. (1989) Repeated antigen challenge induces airway hyperresponsiveness with tissue eosinophilia in guinea pigs. *Journal of Applied Physiology*, **67**, 1133–1139.

KATZ, J., HARMON, C. C., BUCKNER, G. P., RICHARDSON, G. J., RUSSELL, M. W. & MICHALEK, S. M. (1993) Protective salivary immunoglobulin A responses against *Streptococcus mutans* infection after intranasal immunization with *S. mutans* antigen I/II coupled to the B subunit of cholera toxin. *Infection and Immunity*, **61**, 1964–1971.

KAWABATA, T. T. & BABCOCK, L. S. (1993) Measurement of murine ovalbumin-specific IgE by a rat basophil leukemia cell serotonin release assay. Comparison to the rat passive cutaneous anaphylaxis assay. *Journal of Immunological Methods*, **162**, 9–15.

KAWABATA, T. T., BABCOCK, L. S. & HORN, P. A. (1996) Specific IgE and IgG1 responses to Subtilisin Carlsberg (Alcalase) in mice: development of an intratracheal exposure model. *Fundamental and Applied Toxicology*, **29**, 238–243.

KUDO, K., OKUDAIRA, H., MIYAMOTO, T., NAKAGAWA, T. & HORIUCHI, Y. (1978) IgE antibody response to mite antigen in the mouse. Suppression of an established IgE antibody response by chemically modified antigen. *Journal of Allergy and Clinical Immunology*, **61**, 1–9.

KUNG, T. T., STELTS, D. M., ZURCHER, J. A., ADAMS, G. K., EGAN, R. W., KREUTNER, W., WATNICK, A. S., JONES, H. & CHAPMAN, R. W. (1995) Involvement of IL-5 in a murine model of allergic pulmonary inflammation: prophylactic and therapeutic effect of an anti-IL-5 antibody. *American Journal of Respiratory Cell and Molecular Biology*, **13**, 360–365.

LACK, G., RENZ, H., SALOGA, J., BRADLEY, K. L., LOADER, J., LEUNG, D. Y., LARSEN, G. & GELFAND, E. W. (1994) Nebulized but not parenteral IFN-gamma decreases IgE production and normalizes airways function in a murine model of allergen sensitization. *Journal of Immunology*, **152**, 2546–2554.

LEI, H.-Y., LEE, S.-H. & LEIR, S.-H. (1996) Antigen-induced anaphylactic death in mice. *International Archives of Allergy and Applied Immunology*, **109**, 407–412.

MADI, S., GIESSLER, J., HIRSCHELMANN, R., FRIEDRICH, G. & BRAQUET, P. (1991) Allergen-induced bronchospasm in passively sensitized guinea pigs: influence of new substances in comparison to reference compounds. *Agents and Actions*, **32**, 144–145.

MANCINO, D. & OVARY, Z. (1980) Adjuvant effects of amorphous silica and of aluminium hydroxide on IgE and IgG1 antibody production in different inbred mouse strains. *International Archives of Allergy and Applied Immunology*, **61**, 253–258.

MAPP, C., HARTIALA, J., FRICK, O. L., SHIELDS, R. L. & GOLD, W. M. (1985) Airway responsiveness to inhaled antigen, histamine, and methacholine in inbred, ragweed-sensitized dogs. *American Review of Respiratory Disease*, **132**, 292–298.

MARKHAM, R. J. & WILKIE, B. N. (1976) Influence of detergent on aerosol allergic sensitization with enzymes of *Bacillus subtilis*. *International Archives of Allergy and Applied Immunology*, **51**, 529–543.

(1979) Effect of detergent on protease-induced mortality and pulmonary histopathology. *Archives of Environmental Health*, **34**, 418–423.

MARTIN, J. G., XU, L. J., TOH, M. Y., OLIVENSTEIN, R. & POWELL, W. S. (1993) Leukotrienes in bile during the early and the late airway responses after allergen challenge of sensitized rats. *American Review of Respiratory Disease*, **147**, 104–110.

MAUSER, P. J., PITMAN, A. M., FERNANDEZ, X., FORAN, S. K., ADAMS, G. K., KREUTNER, W., EGAN, R. W. & CHAPMAN, R. W. (1995) Effects of an antibody to interleukin-5 in a monkey model of asthma. *American Journal of Respiratory and Critical Care Medicine*, **152**, 467–472.

MCCASKILL, A. C., HOSKING, C. S. & HILL, D. J. (1982) IgE and haemagglutinating antibody production in mice following intranasal immunization with ryegrass pollen. *International Archives of Allergy and Applied Immunology*, **67**, 191–193.

(1984) Anaphylaxis following intranasal challenge of mice sensitized with ovalbumin. *Immunology*, **51**, 669–677.

MINSHALL, E. M., RICCIO, M. M., HERD, C.M., DOUGLAS, G. J., SEEDS, E. A., MCKENNIFF, M. G., SASAKI, M., SPINA, D. & PAGE, C. P. (1993) A novel animal model for investigating persistent airway hyperresponsiveness. *Journal of Pharmacological and Toxicological Methods*, **30**, 177–188.

MISAWA, M. & SUGIYAMA, Y. (1993) An airway hyperresponsiveness model in rat allergic asthma. *Arerugi*, **42**, 107–114.

MITCHELL, G. F. (1976) Studies on immune responses to parasite antigens in mice. II. Aspects of the T cell dependence of circulating reagin production to *Ascaris suum* antigens. *International Archives of Allergy and Applied Immunology*, **52**, 79–94.

MULLER, C. P. & BRONS, N. H. (1994) Systemic B-cell response in the mouse after mucosal immunization with measles virus. *Immunity and Infection*, **22**, 125–126.

NAKASHIMA, S., KAMIKAWA, H. & OGITA, Z. (1983) Isoelectric focusing of the inbred mouse antibody to bacterial alpha-amylase. *Journal of Biochemistry (Tokyo)*, **94**, 1723–1730.

OETTGEN, H. C., MARTIN, T. R., WYNSHAW-BORIS, A., DENG, C., DRAZEN, J. M. & LEDER, P. (1994) Active anaphylaxis in IgE-deficient mice. *Nature*, **370**, 367–370.

OSHIBA, A., HAMELMANN, E., TAKEDA, K., BRADLEY, K. L., LOADER, J. E., LARSEN, G.L. & GELFAND, E. W. (1996) Passive transfer of immediate hypersensitivity and airway hyperresponsiveness by allergen-specific immunoglobulin (Ig)E and IgG1 in mice. *Journal of Clinical Investigation*, **97**, 1398–1408.

PETTY, R. E. & STEWARD, M. W. (1977) The effect of immunological adjuvants on the relative affinity of anti-protein antibodies. *Immunology*, **32**, 49–55.

POIRIER, T. P., KEHOE, M. A. & BEACHEY, E. H. (1988) Protective immunity evoked by oral administration of attenuated aroA *Salmonella typhimurium* expressing cloned streptococcal M protein. *Journal of Experimental Medicine*, **168**, 25–32.

RENZ, H., SMITH, H. R., HENSON, J. E., RAY, B. S., IRVIN, C. G. & GELFAND, E. W. (1992) Aerosolized antigen exposure without adjuvant causes increased IgE production and increased airway responsiveness in the mouse. *Journal of Allergy and Clinical Immunology*, **89**, 1127–1138.

RENZ, H., LACK, G., SALOGA, J., SCHWINZER, R., BRADLEY, K., LOADER, J., KUPFER, A., LARSEN, G. L. & GELFAND, E. W. (1994) Inhibition of IgE production and normalization of airways responsiveness by sensitized CD8 T cells in a mouse model of allergen-induced sensitization. *Journal of Immunology*, **152**, 351–360.

REVOLTELLA, R. & OVARY, Z. (1969) Reaginic antibody production in different mouse strains. *Immunology*, **17**, 45–54.

RITZ, H. L., EVANS, B. L., BRUCE, R. D., FLETCHER, E. R., FISHER, G. L. & SARLO, K. (1993) Respiratory and immunological responses of guinea pigs to enzyme-containing detergents: a comparison of intratracheal and inhalation modes of exposure. *Fundamental and Applied Toxicology*, **21**, 31–37.

ROBINSON, M. K., BABCOCK, L. S., HORN, P. A. & KAWABATA, T. T. (1996) Specific antibody responses to subtilisin carlsberg (Alcalase) in mice: development of an intranasal exposure model. *Fundamental and Applied Toxicology* (in press).

SARLO, K., POLK, J. E., & RITZ, H. L. (1991) Guinea pig intratracheal test to assess respiratory allergenicity of detergent enzymes; comparison with the human data base. *Journal of Allergy and Clinical Immunology*, **87**, 816.

SEHON, L. Z. & MOHAPATRA, S. S. (1992) Induction of IgE antibodies in mice with recombinant grass pollen antigens. *Immunology*, **76**, 158–163.

SIKORSKI, E. E., MCCAY, J. A., WHITE, K. L., JR., BRADLEY, S. G. & MUNSON, A. E. (1989) Immunotoxicity of the semiconductor gallium arsenide in female B6C3F1 mice. *Fundamental and Applied Toxicology*, **13**, 843–858.

STARCHER, B. & WILLIAMS, I. (1989) A method for intratracheal instillation of endotoxin into the lungs of mice. *Laboratory Animals*, **23**, 234–240.

STARUCH, M. J. & WOOD, D. D. (1982) Genetic influences on the adjuvanticity of muramyl dipeptide *in vivo*. *Journal of Immunology*, **128**, 155–160.

TAKAFUJI, S., SUZUKI, S., KOIZUMI, K., TADOKORO, K., MIYAMOTO, T., IKEMORI, R. & MURANAKA, M. (1987) Diesel-exhaust particulates inoculated by the intranasal route have an adjuvant activity for IgE production in mice. *Journal of Allergy and Clinical Immunology*, **79**, 639–645.

TAKAFUJI, S., SUZUKI, S., KOIZUMI, K., TADOKORO, K., OHASHI, H., MURANAKA, M. & MIYAMOTO, T. (1989) Enhancing effect of suspended particulate matter on the IgE antibody production in mice. *International Archives of Allergy and Applied Immunology*, **90**, 1–7.

TARAYRE, J. P., ALIAGA, M., BARBARA, M., TISSEYRE, N., VIEU, S. & TISNE VERSAILLES, J. (1992) Model of bronchial allergic inflammation in the Brown Norway rat. Pharmacological modulation. *International Journal of Immunopharmacology*, **14**, 847–855.

TAYLOR, W. A., SHELDON, D. & FRANCIS, D. H. (1980) IgE antibody formation in BALB/c mice without adjuvant: induction of responses to grass pollen extract and to a hapten–carrier conjugate. *Immunology*, **39**, 583–588.

TOMLINSON, L. A., CHRISTIE, J. F., FRASER, E. M., MCLAUGHLIN, D., MCINTOSH, A. E. & KENNEDY, M. W. (1989) MHC restriction of the antibody repertoire to secretory antigens, and a major allergen, of the nematode parasite *Ascaris*. *Journal of Immunology*, **143**, 2349–2356.

WALSH, E. E. (1994) Humoral, mucosal and cellular immune response to topical immunization with a subunit respiratory syncytial virus vaccine. *Journal of Infectious Diseases*, **170**, 345–350.

WANG, J. M., DENIS, M., FOURNIER, M. & LAVIOLETTE, M. (1994) Experimental allergic bronchopulmonary aspergillosis in the mouse: immunological and histological features. *Scandinavian Journal of Immunology*, **39**, 19–26.

WEGNER, C. D., GUNDEL, R. H., ABRAHAM, W. M., SCHULMAN, E. S., KONTNY, M. J., LAZER, E. S., HOMON, C. A., GRAHAM, A. G., TORCELLINI, C. A. & CLARKE, C. G. (1993) The role of 5-lipoxygenase products in preclinical models of asthma. *Journal of Allergy and Clinical Immunology*, **91**, 917–929.

9

Chemical Respiratory Allergy: a Regulatory Viewpoint

PETER EVANS

Health and Safety Executive, Bootle, Merseyside

9.1 Introduction

As part of their mission to ensure that risks to people's health and safety from work are properly controlled, governmental regulatory authorities have needed to put in place appropriate regimes to control low-molecular-weight chemicals that have the potential to cause respiratory allergy in the workplace. In recent years this issue has been viewed as a high priority, as the toll of cases of asthma due to reactive chemicals has increasingly become apparent. However, many aspects of our present understanding of the processes whereby low-molecular-weight chemicals (as opposed to proteins and other macro-molecular agents) cause respiratory hypersensitivity have served to confound regulators in their efforts. These aspects will be described in this introduction, in order to provide a context for a review of legislation and regulatory approaches to controlling chemical respiratory allergy.

One basic, but seemingly intractable, source of confusion is the shifting and sometimes overlapping definitions available for key terms surrounding this issue, such as hypersensitivity, respiratory sensitization, allergy and asthma. Medical, regulatory, industrial and academic scientists may each have their preferences. This lack of clarity surrounding definitions has been compounded by uncertainties regarding the underlying toxicological mechanisms of the resulting disease. Thus, possible meanings for respiratory hypersensitivity in relation to effects in the lung include asthma induced by a proven immunological mechanism, asthma induced by an immunological mechanism – proven or presumed, asthma induced by specific (immunological or non-immunological) mechanisms or asthma induced by any mechanism. A further possible refinement to these definitions would be to differentiate immunological mechanisms into those mediated by immunoglobulin E (IgE) and those apparently not. Given the current relatively rudimentary state of knowledge concerning mechanisms of asthmagenesis for most low-molecular-weight chemicals, this chapter has a broader scope than respiratory allergy (which implies a known immunological mechanism) and covers respiratory hypersensitivity regardless of any particular underlying mechanism. Despite this, it is acknowledged that in time a significant mechanistic role for the immune system may well be revealed for many low-molecular-weight asthmagens.

A second factor that makes the regulation of agents causing respiratory hypersensitivity less straightforward than that for most other toxicological endpoints is the lack of a fully validated predictive animal test. Although a number of animal methods, particularly those using guinea pig and mouse described earlier in this book, show considerable promise, none has yet attained international regulatory recognition as a test guideline adopted by the Organization for Economic Co-operation and Development (OECD). At present, the available methods are generally considered by regulators to be acceptable as screening tests that provide useful information for chemicals giving positive responses that will lead, for example, to classification as a respiratory sensitizer. However, negative findings produced by these methods are taken not to be a reliable indicator of the absence of respiratory sensitizing potential. The challenge currently facing these methods is to achieve full validation, so that negative findings (which lead to an 'unclassified' status) also become acceptable to regulatory authorities. Clearly, regulators have a vested interest in this validation process, and a responsibility to encourage it. In this context, in the UK the Health and Safety Executive is actively supporting the further validation of the mouse IgE test and cytokine fingerprinting, with a view to increasing their regulatory acceptance as predictors of the respiratory sensitizing potential of chemicals of occupational concern.

Similarly, no standard, validated *in vitro* method is available, although the potential for a chemical to interact with protein can be considered a prerequisite for immunological activity and the induction of allergic sensitization. Indeed, determination *in vitro* of the ability of low-molecular-weight chemicals to conjugate with protein has been proposed as part of a tiered approach to evaluating respiratory allergenicity (Sarlo and Clark, 1992) and the further validation of such a screen would be welcome. Another potential source of information, structure–activity relationship modelling, is still at a relatively early stage of development, although simple examination of molecular structure for reactive groups and checking whether a particular chemical is of a type (isocyanates or anhydrides, for example) already associated with respiratory hypersensitivity is clearly worthwhile.

Given the absence of routinely-used animal and *in vitro* test methods, most and in some cases perhaps all of the information available for an evaluation of the potential of a chemical to cause respiratory hypersensitivity is based on clinical and epidemiological findings in people exposed at the workplace or, occasionally, at home or in other environments. From the regulatory perspective this information can suffer from a number of deficiencies, some deriving from the nature of the original purpose of the investigations. Many studies have been aimed at the clinical diagnosis of asthma in a patient without any particular need to identify stringently the agent responsible for inducing the asthma, as opposed to that provoking the asthmatic symptoms. Exposure data for the period leading up to the recognition of occupational asthma are rarely available, and in many cases unquantified, but possibly high, previous and/or concurrent exposures to chemicals other than the one under suspicion may prevent a firm conclusion being drawn.

The clinical investigations themselves may contribute further uncertainty by the nature of their conduct and the interpretation of their findings. An example is provided by the bronchial challenge test, which is often considered to be the 'gold standard' for the diagnosis of asthma attributable to a specific chemical. For such a test to be considered stringent by regulatory standards, a series of conditions should be met, including use of a clearly sub-irritant concentration of putative allergen, maintaining blind conditions for the subject, and preferably the investigator as well, as to the nature of the exposure (i.e. whether test or control substance) and careful control of possible confounding factors, such as use of asthma medication, smoking and upper respiratory tract viral infection. For a positive result to be convincing, in a regulatory sense, the response should be of an appropriate magnitude (a decrease in FEV_1 of 20 per cent or greater, for example) over and above any effect seen at the control challenge. Unfortunately, it is unusual for bronchial challenge tests reported in the scientific literature to meet all or even most of these conditions. Thus, it often becomes necessary to take a balanced view of all the information available in order to make the best scientific judgment possible.

Another uncertain aspect of the regulatory control of agents causing respiratory hypersensitivity concerns the importance of peak exposures in the induction of the asthmatic state. Such peaks, consisting of brief periods (perhaps of less than a minute) of exposure to high concentrations of chemical, may be masked in 8-hour time-weighted average values derived by personal sampling, but in fact reflect the intermittent nature of exposures in many industrial processes. At present, practical experience in several industries suggests that peak exposures are significant to the induction of a sensitized state, although the scientific evidence remains inconclusive (Morris, 1994).

There is also for chemicals causing respiratory hypersensitivity some uncertainty regarding the relevant routes of exposure for the induction as opposed to the provocation phase of sensitization. For the latter phase, clearly the inhalation route is the only relevant one. For protein and other macromolecular allergens, it is likely that inhalation is also the only route involved at the induction phase, as effective skin penetration is unlikely. For low-molecular-weight chemicals, however, there is some evidence, largely from animal studies, that an immune response sufficient to sensitize the respiratory tract may appear after dermal exposure (Kimber and Wilks, 1995). As is pointed out, this is unsurprising for an underlying immunological mechanism, since IgE antibody responses are systemic in nature. The current regulatory view, however, while accepting that for a limited number of chemicals there is some evidence from animal experiments that respiratory hypersensitivity can be induced by skin contact, doubts the general relevance of these findings to humans. By this view, any regulatory reference to respiratory hypersensitivity being induced by dermal exposure must await further evidence that such induction can occur in humans.

Against this background of ill-defined toxicological mechanisms, inadequate experimental test systems, inconclusive clinical data and uncertainty regarding the toxicological significance of different exposure patterns and routes, perhaps understandably the development of regulatory controls and guidance aimed specifically at low-molecular-weight chemicals producing respiratory hypersensitivity has encountered some difficulties along the way. Only in the last year, for example, have any meaningful criteria emerged within the European Union (EU) to aid in the classification of chemicals with respect to their potential to cause respiratory hypersensitivity. It is also the case that only in relatively recent years has the full extent of occupational asthma due to low-molecular-weight chemicals become apparent, a process assisted in the UK by the introduction of schemes such as the Surveillance of Work-related and Occupational Respiratory Disease (SWORD), and by additions to the Department of Social Security list of agents for which occupational asthma is a prescribed disease. It is now clear that occupational asthma has become the major cause of work-related respiratory ill-health in the UK, a phenomenon due, among other factors, to a decline in more traditional diseases such as bronchitis, byssinosis and pneumoconiosis and to the increased industrial use of reactive chemicals. It is this realization that has prompted the recent heightened regulatory activity that is described below in sections considering classification and labelling, occupational exposure limits and workplace risk assessment, and supply-side programmes involving notification and risk assessment of new and existing chemicals.

9.2 Hazard Identification, Classification and Labelling

A crucial starting point within the UK/EU framework for regulation of industrial chemicals is identification of the hazardous properties of substances. The classification system in place in the EU serves to identify the hazardous properties of industrial chemicals that are supplied commercially, and is a statutory requirement within each of the member states. Criteria used to derive the appropriate classification and labelling for a substance are available in Annex VI to the original directive (the Dangerous Substances Directive, 67/548/EEC; EEC, 1967), an annex commonly referred to as the 'labelling guide' (93/21/EEC; EEC, 1993a). In the UK, these EU requirements are currently implemented by the Chemicals (Hazard Information and Packaging for Supply) Amendment Regulations 1996, commonly known as 'CHIP 96' (HSE, 1996a), and an 'approved guide' containing the EU criteria is available (HSE, 1994a).

At the moment, the guidance given in the EU labelling guide with respect to sensitization by inhalation is not particularly illuminating. It is stated that substances (and preparations) should be classified in the category of danger 'sensitizing' and assigned the symbol Xn, with the indication of danger 'harmful' and the working phrase R42 (May cause sensitization by inhalation), if at least one of the following two criteria apply:

1 If practical evidence is available which shows the substances and preparations to be capable of inducing a sensitization reaction in humans by inhalation, at a greater frequency than would be expected from the response of a general population.

2 If the substance or preparation is an isocyanate, unless there is evidence that the substance or preparation does not cause sensitization by inhalation.

As a result of the application of these criteria, a total of approximately 20 individual substances with R42 were included in Annex I to the Dangerous Substances Directive. This annex is a compilation of several thousand agreed classification and labelling entries and is represented in the UK by the Approved Supply List (HSE, 1996b). These R42 substances included isocyanates, anhydrides, cobalt, nickel sulphate, butadiene epoxide, glycidol, methenamine and several complex dyes. For a few of these substances the toxicological evidence underpinning their receipt of R42 is not obvious, and when confronted with the need to classify chemicals with considerably more complex databases, such as glutaraldehyde, ethylenediamine and methyl methacrylate, the initiative was taken in 1993 to develop more detailed criteria for respiratory hypersensitivity. The aim was to make assignment of R42 a more consistent, transparent and scientifically-based process, and the revised criteria were formally adopted by EU member states in May 1996, with the intention that they came into effect in national law by 31 May 1998. In the UK, the criteria will be incorporated in a revision of the CHIP Regulations in 1998. The revised criteria for assignment of R42 are as follows.

1 If there is evidence that the substance or preparation can induce specific respiratory hypersensitivity.

2 Where there are positive results from appropriate animal tests.

3 If the substance is an isocyanate, unless there is evidence that the substance does not cause respiratory hypersensitivity.

A series of comments follows the criteria to aid in their interpretation. It is noted that the respiratory hypersensitivity in question is normally seen as asthma, but that 'other hypersensitivity reactions such as rhinitis and alveolitis are also considered'. Of particular significance is the statement that although 'the condition will have the clinical character of an allergic reaction . . . immunological mechanisms do not have to be demonstrated'. There is also a key comment that starts to explain the distinction between the identification of a chemical as having the innate ability to cause respiratory hypersensitivity (perhaps in only a very small proportion of those exposed) and the need to classify it as a sensitizer and apply R42. This is that the decision on classification needs to take into account 'the size of the population exposed' and 'the extent of the exposure'. The implication is that there needs to be a substantial number of cases of asthma induced by a particular chemical in relation to the total number of people exposed to it, before classification becomes appropriate. Thus, a high-production-volume chemical, found in large quantities in many workplaces throughout the world, might not warrant the R42 phrase if only a few cases of asthma associated with its use have been reported over the years. In contrast, 3 cases of asthma among a workforce of 20 in contact with a speciality chemical might well indicate the need for classification as a sensitizer. This sort of 'clustering' of cases can provide strong evidence with respect to a particular chemical. Clearly, however, the cases would need careful clinical investigation and the possibility of shared exposure with another, unsuspected agent (chemical or biological) should also be considered. Although, then, such clustering is considered significant, setting an arbitrary asthma prevalence threshold of say 5 per cent of the workforce exposed to a particular chemical, above which R42 could be routinely applied, was not considered scientifically sound. Thus, prevalence information needs to be examined for each chemical on a case-by-case basis in order to achieve a more reliable assessment.

The first criterion for R42 refers to evidence for induction of specific respiratory hypersensitivity. The accompanying comments explain that suitable evidence can come in two forms. One form, considered to be adequate on its own, consists of data from positive bronchial challenge tests conducted according to accepted guidelines. The plural 'tests' implies that at least two are required, while 'accepted guidelines' refers to any standardized, preferably published, protocol that reflects good practice. Although positive bronchial challenge tests are considered to provide sufficient evidence for classification on their own, it is recognized that in practice many other investigations related to clinical history and lung function will already have been carried out.

The second type of evidence is clinical history and appropriate lung function data related to the chemical, together with supportive information. This supportive evidence can include structure–activity considerations, *in vivo* and *in vitro* immunological test results, but also the findings of mechanistic studies indicating that the hypersensitivity is mediated by effects that appear to be non-immunological. Two examples given of possible non-immunological mechanisms are pharmacological and repeated low-level irritation. With regard to the latter, it is only recently that any substantive supporting evidence has begun to become available (Kipen *et al.*, 1994). It is emphasized that 'substances that only elicit symptoms of asthma by irritation in people with bronchial hyperreactivity should not be assigned R42'. This means that substances are not classified if they only trigger asthmatic responses in individuals with airways rendered hyperresponsive by other, pre-existing conditions, such as congenital asthma or bronchitis.

The second criterion for R42 refers to positive results from appropriate animal tests. A brief comment adds that suitable animal data may include IgE measurements (for example, in mice) and specific pulmonary responses in guinea pigs. This comment has been deliberately left rather vague, in order to accommodate future developments and refinements in the area, but it is clear that the 'appropriate' methods to emerge are likely to be based on existing protocols (for example Karol et al., 1985; Dearman et al., 1992).

The third criterion for R42 maintains the regulators' simple presumption that any chemical containing the free isocyanate group may cause respiratory hypersensitivity – this seems sensible, since isocyanates routinely head the annual lists for new cases of occupational asthma in the UK (HSE, 1995a).

Under the EU scheme, chemicals which are classified as dangerous to health have appropriate standard 'safety' or S-phrases assigned to their labels, and included in their safety data sheets, to warn users to take particular precautions. For solid substances and preparations which have R42, S22 ('Do not breathe dust') is obligatory on the label.

A further feature of the EU scheme is the requirement to classify and label preparations (intentional mixtures of individual chemicals) according to the Dangerous Preparations Directive (88/379/EEC; EEC, 1988) and its series of adaptations to technical progress, the most recent being the third (93/18/EEC; EEC, 1993b). In the UK this requirement is currently implemented by the CHIP 96 Regulations, and for respiratory hypersensitivity it is relatively straightforward. Preparations other than gaseous should be classified as sensitizing, and assigned the symbol Xn and the phrase R42, if they contain a substance already assigned R42 at a concentration of 1 per cent or greater. For gaseous preparations, the concentration limit is 0.2 per cent (volume/volume) rather than 1 per cent. Under the Dangerous Preparations Directive, for some categories of danger, including sensitizing, there is potentially the option to test the preparation itself in order to determine whether classification is warranted. However, for respiratory hypersensitivity this is currently not a practical option, given the state of validation of the available animal tests.

The use of 1 per cent as the concentration limit for classification of non-gaseous preparations (or 0.2 per cent for gaseous preparations) with respect to respiratory hypersensitivity is somewhat arbitrary, and it is feasible that some relatively potent substances present at levels below this limit may still be able to induce asthma. In these circumstances, scientific evidence that the standard concentration limit (i.e. 1 per cent) is inappropriate for protective purposes can be used to set a limit that is specific to the chemical – there is space in the Annex 1 entry for each substance for any such specific concentration limit.

There may also be concern that subjects who are already sensitized would be unlikely to be protected by a 1 per cent limit from an asthmatic reaction provoked by the substance, since it is thought that provocation can occur at much lower concentrations than induction. One solution might be to have two distinct concentration limits for each respiratory allergen, with that for provocation being perhaps 100- or even 1000-fold lower than that for induction. However, given the very sparse information generally available regarding thresholds for both of these processes, this seems an over-elaborate approach. Perhaps a more practical proposal would be for the identity of a sensitizer to be declared on the label of a preparation when it is present at a concentration below the limit for assigning R42, as a warning for those who are already sensitized to the substance. It should be noted, though, that these and other possible approaches are still at an early stage of consideration within the EU.

Compared with developments in the EU, legislation and criteria pertaining to classification of chemicals causing respiratory hypersensitivity in other OECD countries appears rather rudimentary, other than in Norway, which is in any case intending to implement the revised EU criteria. Australia relies on EU criteria, while Japan apparently has no legislation or criteria in relation to such substances. The Canadian Controlled Products Regulations currently define respiratory tract sensitization as 'the development in a person who is not atopic of severe asthma-like symptoms on exposure to a substance to which the person has been exposed'. A substance or preparation is classified as a 'respiratory tract sensitizer' if there is evidence that it causes 'respiratory tract sensitization in persons following exposure to it in the workplace'. Canada has a concentration limit of 0.1 per cent for use when classifying preparations in the absence of test results. In the USA, the Occupational Safety and Health Administration (OSHA) has a general definition for sensitizer: 'a chemical that causes a substantial proportion of exposed people or animals to develop an allergic reaction in normal tissue after repeated exposure to the chemical'. The concentration limit applied to preparations in the absence of test results is 1 per cent, the same as in the EU. In the OSHA scheme, however, labelling is also required if the chemical is present in the preparation at less than 1 per cent but can be expected to be released, thereby exceeding OSHA permissible exposure limit or the American Conference of Governmental Industrial Hygienists (ACGIH) threshold limit values.

9.3 Risk Assessment, Occupational Exposure Limits and the UK COSHH Regulations

Setting standards for airborne exposure to chemicals in workplace air is a key element of occupational risk management and is relevant to the control of agents producing respiratory hypersensitivity, as it is to substances with other toxicological properties. In Great Britain, occupational exposure limits have legal force provided by the Control of Substances Hazardous to Health

(COSHH) Regulations 1994 (HSE, 1994b). Similar measures apply in Northern Ireland. There are two types of limit, with the defining difference being that an Occupational Exposure Standard (OES) is set at a level for which there is no evidence of injury to health, while for a Maximum Exposure Limit (MEL) such a level cannot be identified or, if it can, socio-economic factors indicate that control to that level is not reasonably practicable across the breadth of industrial uses for the chemical. The term 'not reasonably practicable' implies that the effort and cost of any measures required to control the exposure level in question would be grossly disproportionate to the benefit obtained.

An OES is 'the concentration of an airborne substance, averaged over a reference period, at which, according to current knowledge, there is no evidence that it is likely to be injurious to employees if they are exposed by inhalation, day after day, to that concentration' (HSE, 1996c). This is therefore a 'health-based' standard and control of exposure to this level is deemed to be adequate under COSHH. The process of deciding its numerical value for a particular chemical involves the determination, for the critical health effect, of a no-observed-adverse-effect level, preferably one based on human findings and inhalation exposure. Often, however, there are available only data from studies in animals and/or using routes of exposure other than the relevant one, and compensation for the resulting uncertainty may be necessary using an appropriate factor. In comparison, a MEL is 'the maximum concentration of an airborne substance, averaged over a reference period, to which employees may be exposed by inhalation under any circumstances'. The MEL is not a 'health-based' standard and consequently there is a duty imposed by COSHH for the employer to ensure that exposure is kept as far below the MEL as is reasonably practicable.

Guidance on matters such as the criteria used to decide which of the two types of limit is applicable, together with lists of prevailing values for the limits for a large number of substances, is published annually by the regulatory authority (HSE, 1996c). Following the general principles of COSHH, this guidance states that exposure to respiratory sensitizers should be prevented, wherever this is reasonably practicable. If prevention of exposure is not possible, 'the primary aim must be to prevent employees from developing respiratory sensitization by assessing the risks and applying adequate standards of control'. Where there is limited or no information to allow in-house exposure limits to be set, exposure should be reduced as low as is reasonably practicable. It is also noted that activities giving rise to peak concentrations over short periods should receive particular attention, and that contamination of other working areas should be minimized. Finally, it is considered that 'health surveillance is appropriate for all employees exposed to respiratory sensitizers unless the COSHH assessment has shown that there is unlikely to be a risk of sensitization under the conditions of use'.

Some of the substances in the published lists of limits have been assigned what is referred to as the 'Sen notation', meaning that they are 'capable of causing respiratory sensitization'. The identified substances are those in the

Approved Supply List which have the R42 phrase (May cause sensitization by inhalation), or are listed under legislation associated with prescribed or reportable diseases. It perhaps says something about the uncertainty still surrounding so many aspects of the control of respiratory sensitizers that even in 1996 the list of occupational exposure limits contains few substances with the Sen notation. The recent heightened regulatory effort in the area of respiratory sensitizers does mean, however, that a number of other chemicals, for which concern has arisen regarding induction of asthma in the workplace, have been or are being reviewed as part of the limit-setting process. The general experience in reviewing the toxicological data on respiratory sensitizers is that the information available precludes the establishment of a clear no-adverse-effect-level for this endpoint. Hence confirmed respiratory sensitizers are usually allocated a MEL within the UK system.

Although the EU has been attempting for some years to introduce a system of occupational exposure limits, its current status remains somewhat embryonic and ill-defined. A clear strategy and set of principles for dealing with substances producing respiratory hypersensitivity have not yet emerged.

In the UK, occupational exposure limits are part of the general framework of the COSHH Regulations 1994, which set out the legal requirements for employers in attempting to protect people in the workplace against the effects of hazardous chemicals including agents causing respiratory hypersensitivity. The regulatory authority has published an Approved Code of Practice, which gives practical guidance on the Regulations (HSE, 1995b). Such a Code has a special legal status in that if an employer is prosecuted and shown not to have followed a relevant provision of the Code, it will be found at fault unless it can show that it complied with the law in some other way.

There is also specific guidance available for respiratory sensitizers in regard to compliance with COSHH (HSE, 1994c). This guidance examines the relevant COSHH Regulations, particularly Regulation 6 (assessment of health risks), 7 (prevention or control of exposures), 8 (use of control measures), 9 (maintenance, examination and test of control measures), 10 (monitoring exposure at the workplace), 11 (health surveillance) and 12 (information, instruction and training), and provides practical advice in the form of check-lists and case studies.

The document also includes two lists of chemicals, the first comprising substances responsible for most cases of occupational asthma in the UK, as indicated by the findings of the SWORD surveillance scheme (Meredith and McDonald, 1994). The second list gives some other substances that have been reported in the scientific and medical literature to cause respiratory sensitization. It has, however, become evident that, particularly for substances in this second list, there is a need to assess critically the toxicological information underpinning their categorization as respiratory sensitizers. Such an assessment is best carried out against accepted criteria, such as those already described that have been developed by the EU for hazard identification and classification purposes. This process of critical appraisal has proceeded in the

Chemical Respiratory Allergy: a Regulatory Viewpoint

UK within regulatory programmes associated with the classification of industrial chemicals and the setting of occupational exposure limits, and it is intended to publish the outcomes in the form of a 'compendium of occupational asthmagens'. This compendium will provide critical assessments of agents implicated in the induction of occupational asthma.

9.4 Supply-side Notification and Risk Assessment Programmes for New Substances

In 1979, a directive (79/831/EEC; EEC, 1979) established for what was then the European Community (EC) a notification scheme for new substances, which are defined as chemicals not on the EC market between January 1971 and September 1981. Before placing a new substance on the market, the manufacturer or importer must notify it in the appropriate member state. Notification involves, among other things, provision of a technical dossier describing various toxicological tests conducted using the substance, and a classification proposal based on the hazards identified by those tests. In the UK, the directive is implemented by the Notification of New Substances Regulations 1993 (HSE, 1993).

For a new substance that is to be initially supplied at the 'base-set' level (greater than 1 tonne per annum), information regarding acute and repeated-dose toxicity, skin and eye irritation, skin sensitization and genotoxicity is necessary. Testing to identify any respiratory sensitization hazard is not required, although structure–activity considerations may lead to classification as sensitizing and assignment of R42 (May cause sensitization by inhalation), as described previously for isocyanates. There is also the option to apply the safety phrase S22 ('Do not breathe dust') in the absence of R42, and this has been done, for example, with certain new, dusty chlorotriazinyl dyestuffs.

In 1992, the so-called 'seventh amendment' (92/32/EEC; EEC, 1992) to the original Dangerous Substances Directive developed the notification process further, and in particular brought in a requirement for the 'real or potential risk to man and the environment' to be assessed. Legally binding principles for this process were subsequently published in a separate 'risk assessment' directive (93/67/EEC; EEC, 1993c). Supplementary guidance, which is not legally binding, on more detailed, technical aspects of risk assessment is also available throughout the EU (HSE, 1994d).

This guidance describes in general terms each of the four elements (hazard identification, dose [concentration]–response [effect] assessment, exposure assessment and risk characterization) that together comprise risk assessment. For a number of toxicological endpoints, including reproductive toxicity, mutagenicity and carcinogenicity, there is also a considerable amount of specific technical guidance, including detailed testing strategies for use as the supply tonnage of the chemical increases. In contrast, the guidance that it is presently possible to provide in relation to respiratory hypersensitivity is rudimentary. It is noted that there is no accepted test method for respiratory sensitization, that definition of a no-observed-adverse-effect-level is not possible, and that there is currently no consensus on the possibility of identifying a provocation concentration for a sensitizing substance, below which adverse effects are unlikely to occur in people who are already sensitized. Regarding the risk characterization stage, it is stated that if 'respiratory sensitization of people to a substance does occur, adverse effects, which may be life-threatening, may occur in sensitized individuals following exposure to extremely low amounts/concentrations of the substance. Sensitized individuals may also be sensitive to other, chemically related substances. Particular precautions are necessary to protect such people from exposure to the substance(s) of concern and recommendations for risk reduction for sensitizers may need to take this possibility into account'. Clearly, guidance that is much more informative than this is desirable for a toxicological endpoint with such potential significance to human health as respiratory allergy. However, the ability to produce such guidance awaits an improved understanding in many of the areas of uncertainty already described. At present, it is difficult to see how any immediate progress can be made.

A different sort of notification scheme operates for new chemicals in the USA, where companies are required under the Toxic Substances Control Act to file with the Environmental Protection Agency (EPA) a 'premanufacture notice'. This they should do 90 days before they manufacture or import any industrial chemical not on an EPA list of commercial chemicals. In contrast to the EU notification scheme, companies are not expected to submit test data on the chemical when filing a premanufacture notice. Instead, the EPA screens the new substance by comparing it with structurally similar chemicals; in a proportion of cases it subsequently imposes testing requirements. Among structures considered suspect in relation to respiratory hypersensitivity by the EPA are isocyanates and anhydrides, and testing for respiratory sensitizing potential has been required for such chemicals as part of the premanufacturing notice programme. Historically, a method using guinea pigs has been recommended for this testing (Karol *et al.*, 1985). More recently, however, the general applicability of this procedure has been questioned, and the options for other approaches, including screening with the mouse IgE test, have been considered (Selgrade *et al.*, 1994).

9.5 Supply-side Risk Assessment Programmes for Existing Substances

Within the EU, an 'existing substance' is one of the 110 000 or so chemicals on the EC market at some time between January 1971 and September 1981. Such substances by definition have not gone through the notification procedure that was instigated at the end of this period, and the quality of the toxicological and other data available for their risk assessment is very variable. This situation has been addressed by the Existing Substances Regulation (EEC, 1993d), which provides for the collection of data on substances produced at high tonnages, prioritization, then risk assessment and identification of any necessary risk reduction measures for the priority substances.

A detailed technical guidance document is available to support the Regulation (EEC, 1994), although, as for the new substances guidance, the information it provides with respect to respiratory hypersensitivity is limited by the surrounding uncertainties. It is again noted that there is no standard test for respiratory sensitization, but adds that 'potentially useful test methods are under active research and development'. It refers to 'mouse IgE or guinea pig models' as providing data which may already be available for the risk assessment of an existing substance. Human information that may be available includes case reports or epidemiological studies reporting 'respiratory (allergic rhinitis, alveolitis or asthma) reactions', and bronchial challenge test results. It is noted that studies with negative findings should also be evaluated, and that 'structural analogies with known sensitizers' may also be considered.

Evaluation criteria for available human data include the number of well-documented cases in relation to the size of the exposed population, and various exposure-related factors – the physical state and concentration/quantity of the substance as well as the frequency and duration of the exposures. The particular point is also made that there may be significant uncertainty in the evaluation of human sensitization data, due to 'poor reporting, lack of specific information on exposure, small number of subjects and concomitant exposure to other chemicals'. In relation to the evaluation of animal data, it is noted that 'no methods are yet fully validated, although predictive methods are promising'. Consideration of the findings of skin sensitization tests in animals is also warranted, since 'it is probable that most, if not all, chemical respiratory allergens also have the potential to cause skin sensitization in experimental models'. However, it is also pointed out that many chemicals that give positive responses in predictive tests for skin sensitization have not been found to induce respiratory sensitization in humans. In relation to the evaluation of any other available information, it is noted that 'some physico-chemical characteristics and biological properties (e.g. reactivity with proteins) appear important correlates of respiratory sensitization'.

Once again it is acknowledged that there can be difficulty in carrying out hazard identification and risk assessment for respiratory hypersensitivity. Although identification of a substance as a respiratory sensitizer is likely to be based on positive findings in humans, it is noted that the exposure conditions that lead to such sensitization are not clearly understood. However, for the risk characterization, the exposure conditions of the studies providing the evidence for respiratory sensitization in humans can be compared with those of the populations (workers, consumers and/or those exposed indirectly via the environment) being considered. When no suitable human evidence is available, the characterization will have to be more pragmatic, taking into account the exposure conditions under which the chemical is used without induction of respiratory hypersensitivity.

Within the context of the Existing Substances Regulation, one of the three possible overall outcomes of the risk assessment procedure for existing substances is that 'there is need for further information and/or testing'. In the case of respiratory sensitization, the guidance states that it is not expected that (further) testing in animals would be justified (reflecting the non-availability of an internationally accepted test guideline), but it is noted that *in vitro* testing (such as investigation of protein binding) and/or further exposure information may sometimes be desirable in order to refine the risk assessment.

9.6 Concluding Remarks

Respiratory sensitization is currently proving to be one of the more difficult and controversial areas of regulatory toxicology. We are hampered by poor existing data, a lack of understanding of underlying toxicological processes, and the absence of an internationally accepted and convenient experimental animal model. It is hoped that over the next few years things will get better!

Acknowledgements

I am grateful to colleagues in HSE's Toxicology Unit, particularly Celia Elliott-Minty, Steve Fairhurst and Derek James, for support and helpful comments.

References

DEARMAN, R. J., BASKETTER, D. A. & KIMBER, I. (1992) Variable effects of chemical allergens on serum IgE concentration in mice. Preliminary evaluation of a novel approach to the identification of respiratory sensitizers. *Journal of Applied Toxicology*, **12**, 317–323.

EEC (1967) Council Directive 67/548/EEC. *Official Journal of the European Communities*, **196**, 1.

—— (1979) Council Directive 79/831/EEC. *Official Journal of the European Communities*, **L259**, 10.

—— (1988) Council Directive 88/379/EEC, *Official Journal of the European Communities*, **L187**, 14–30.

—— (1992) Council Directive 92/32/EEC. *Official Journal of the European Communities*, **L154**, 1–29.

—— (1993a) Annex to Commission Directive 93/21/EEC. *Official Journal of the European Communities*, **L110A**, 45–86.

—— (1993b) Commission Directive 93/18/EEC. *Official Journal of the European Communities*, **L104**, 46.

(1993c) Commission Directive 93/67/EEC. *Official Journal of the European Communities*, **L227**, 9–18.
(1993d) Council Regulation (EEC) No 793/93. *Official Journal of the European Communities*, **L84**, 1–75.
(1994) *Risk assessment of existing substances. Technical guidance document*, **XI/919/94**. Brussels: European Commission DGXI.
HSE (1993) *Notification of New Substances Regulations 1993*, **SI 1993/3050**. London: HMSO.
(1994a) *Approved guide to the classification and labelling of substances and preparations dangerous for supply*, 2nd Edn. Sudbury: HSE Books.
(1994b) *Control of Substances Hazardous to Health Regulations 1994*, **SI 1994/3246**. London: HMSO.
(1994c) *Preventing asthma at work. How to control respiratory sensitisers.* Sudbury: HSE Books.
(1994d) *Risk assessment of notified new substances. Technical guidance document.* Sudbury: HSE Books.
(1995a) *Health and Safety Statistics 1994/95*, pp. 144–145. Sudbury: HSE Books.
(1995b) *General COSHH ACOP*. Sudbury: HSE Books.
(1996a) *Chemicals (Hazard Information and Packaging for Supply) Amendment Regulations 1996*, **SI 1996/1092**. London: HMSO.
(1996b) *Approved Supply List*, 3rd Edn. Sudbury: HSE Books.
(1996c) *EH40/96, Occupational exposure limits 1996*. Sudbury: HSE Books.
KAROL, M. H., STADLER, J. & MAGRENI, C. M. (1985) Immunotoxicologic evaluation of the respiratory system: animal models for immediate- and delayed-onset pulmonary hypersensitivity. *Fundamental and Applied Toxicology*, **5**, 459–472.
KIMBER, I. & WILKS, M. F. (1995) Chemical respiratory allergy. Toxicological and occupational health issues. *Human and Experimental Toxicology*, **14**, 735–736.
KIPEN, H. M., BLUME, R. & HUTT, D. (1994) Asthma experience in an occupational and environmental medicine clinic. Low-dose reactive airways dysfunction syndrome. *Journal of Occupational Medicine*, **36**, 1133–1137.
MEREDITH, S. K. & MCDONALD, J. C. (1994) Work-related respiratory disease in the United Kingdom, 1989–1992: report of the SWORD project. *Occupational Medicine*, **44**, 183–189.
MORRIS, L. (1994) Respiratory sensitisers. Controlling peak exposures. *Toxic Substances Bulletin*, **24**, 10.
SARLO, K. & CLARK, E. D. (1992) A tier approach for evaluating the respiratory allergenicity of low molecular weight chemicals. *Fundamental and Applied Toxicology*, **18**, 107–114.
SELGRADE, M. K., ZEISS, C. R., KAROL, M. H., SARLO, K., KIMBER, I., TEPPER, J. S. & HENRY, M. C. (1994) Workshop on status of test methods for assessing potential of chemicals to induce respiratory allergic reactions. *Inhalation Toxicology*, **6**, 303–319.

Index

acid anhydrides 44, 52
   occupational asthma 1, 31, 37–9, 62
acrylates 47
active cutaneous anaphylaxis test 114
acute pulmonary oedema 1, 30
adenocarcinoma 45
adult respiratory distress syndrome (ARDS) 45
airstream helmets 51
albumin 42, 43, 111
Alcalase 110–1, 116, 136–44
allergic contact dermatitis 3, 73, 82
   mouse models 122, 125
aluminium 46
alveolar haemorrhage 42
alveolar macrophages 40–1, 84, 88, 89
alveolar pneumonitides 52
alveolitis 1, 30, 42, 73, 155, 163
amines 42, 46, 85
aminoethyl ethanolamine 36
amprolium HCl 36
amylase 10, 142, 143, 144
anaphylatoxins 38
anaphylaxis 8, 108
anhydrides 34, 112, 152, 155, 162
   see also acid anhydrides
anthraquinone dyes 46
antibiotics 11
anticholinesterase 38
antigen presentation 37, 83–5, 91, 136
antihistamine 115
Approved Supply List 155, 160
aspirin 21
asthma 1–3, 7–22, 29–40, 44–7, 73–91
   defined 8–9

diagnosis 49–51
guinea pig models 107, 108, 111, 116, 121
history 7–10
isocyanates 122
LMW agents 34–6
mouse models 121, 122, 135, 136
prevention 52, 65–8
regulations 151–8, 160–1, 163
risk assessment and management 61–9
treatment 20–2, 51–2, 153
atopy 10–11, 17, 46, 51, 52, 158
   asthma 31, 64–5
   dermatitis 11
   sensitization 81, 82, 87
autocoids 14, 17
automobile industry 32, 34, 43
azodicarbonamide 36

B lymphocytes 17, 18, 75, 76, 90
B7 proteins 77
basal cells 19
basement membrane 9, 10, 40–1
basic fibroblast growth factor (bFGF) 20
basic proteins 18–20, 85
basophils 12, 19
benign pleural disease 30
berylliosis and beryllium 32, 38, 42–3, 51, 73
biopsies 3, 13–5, 40, 42, 81
bone marrow 13, 76, 86, 87
bronchial challenge test 153, 156, 163
bronchiectasis 51
bronchioles 42
bronchiolitis obliterans 44, 49
bronchitis 30, 44, 45, 154, 157

167

Index

bronchoalveolar lavage (BAL) 9, 12–3, 15, 18, 37, 41, 42
  berylliosis 43
  guinea pig models 111
  sensitization 78, 81, 85, 86
bronchoconstriction 8, 11, 12–3, 18–9, 33, 74, 85
bronchodilators 7–8, 10, 13, 20–1, 49
bronchospasms 8, 9, 17, 40, 41
Brown Norway rats 83, 127
building related illness 30, 48–9
butadiene epoxide 155
byssinosis 30, 33, 39, 154

cadmium 46
caffeine 8
calcitonin gene-related peptide (CGRP) 39
cancer 30, 63
capsaicin 115
carbon dioxide 48
carmine 2
carrier proteins 33, 111, 115
cationic toxins 85
cell adhesion molecules 14–5, 17
cell-mediated immune responses 33, 37, 38, 45, 51, 74–6
cellulase 144
cephalosporin 11, 36
chemokines 86, 87, 91
chloramine-T 2, 11, 46
chlorine gas 45
chlorotriazinyl dyestuffs 161
chromium and chromium salts 11, 35, 45
chromosomes 10, 17
chronic obstructive lung disease 49
chymotrypsin-like enzyme 10
classification of hazards 154–8
cobalt 35, 37, 45–6, 155
collagen 9, 10, 20, 85
colophony 1, 31, 36, 46
columnar cells 19
conducting airways 39–42
conjunctivitis 45, 61, 107
contact exposure and sensitization 73–4, 80–3, 91, 116
  guinea pig models 114–5
  mouse models 122–5, 143
  regulation 153, 154
corticosteroids 10, 18, 21–2, 52
  side effects 21
COSHH Regulations 158–61
costimulation 77
Creola bodies 19
cromolyn sodium 52
cyclosporin A 21
cysteine protease 10
cysteinyl-leukotrienes 12–3, 18–9
cytokine fingerprinting 124, 129–31, 152
cytokine secretion profiles 123–4, 129–31
cytokines 14–15, 16, 17–20, 37

mouse protein studies 136
  occupational asthma 38, 39, 41
  sensitization 75–80, 82, 86–8, 89, 91
cytostimulants 85
cytotoxic immune response 37

decongestants 7
delayed-type reactions 123
  see also late asthmatic response (LAR)
dendritic cells 12, 17, 18, 78, 83–4, 88
dermal exposure *see* contact exposure and sensitization
dermatitis 11, 45
desmosomes 19–20
detergent matrix 138–43
detergent enzymes *see* enzymes
diazonium salts 36, 47
dicyclohexylmethane-4,4'-diisocyanate (HMDI) 114
diesel exhaust particles 89–90
dimethylethanolamine 36
2,4-dinitrochlorobenzene (DNCB) 80, 113, 116
  mouse models 122–7, 129, 130
2,4-dinitrofluorobenzene (DNFB) 124, 129
diphenylmethane diisocyanate 2, 34, 38, 42, 43, 80
  guinea pig models 112, 113, 114, 115
  mouse model 124
dual asthmatic reaction (DAR) 40, 45, 107
dyes 2, 36, 44, 47, 161
  guinea pig models 112, 114, 115, 116
dyspnoea 37

E-selectin 14, 15
early asthmatic response 12–3, 40
eczema 8
emphysema 30, 45
endolethium 14–15
endothelin-1 (ET-1) 20
endotoxins 49
enzyme-linked immunosorbent assay (ELISA) 44
  guinea pig models 108, 110, 112–4
  mouse models 124, 126, 137–40, 142, 143
enzymes 1, 10, 11, 61, 63, 65, 69
  guinea pig models 109–11, 116
  mouse models 135–45
eosinophil apoptosis 13
eosinophil cationic protein (ECP) 20, 85
eosinophil-derived neurotoxin 85
eosinophil infiltration 9, 10, 108, 115
eosinophil peroxidase (EPO) 20, 85
eosinophilia 9, 13, 15–8, 22, 86
eosinophilopoietic cytokines 13
eosinophils 12–20, 22, 33, 37, 41
  sensitization 79, 81, 85–7
eotaxin 87
Ephedra plant 7
ephedrine 7

Index

epithelium 12, 16, 19–20
    damage 9, 10, 19
    sensitization 84, 85, 88, 89
epoxy resin 34, 44
ethylene diamine 36, 155
European Union regulations 154–5, 157–8, 160–2
Evans blue dye 115
extravasation 115

farm workers 29, 31, 32, 42
farmer's lung *see* hypersensitivity pneumonitis
febrile response 109
fibroblasts 12, 20
fibrosis 29, 30, 38, 42–3, 46, 51, 85
    diagnosis 50–1
fishing industry 29
flour 1, 62, 68
fluticasone 21
fluxes 35
food industry 32, 46
formaldehyde 36, 47, 88–9
formoterol 21
freon 36, 47
fungicides 44
furan 47
furfuryl alcohol 36

gelatinase A 20
genetic background 10–2, 64–5, 66, 87, 89, 90–1
glucocorticosteroids 20–1
glutaraldehyde 1, 2, 47, 155
glycidol 155
goblet cells 9, 10, 40
Goodpasture-like syndrome 37
granulocyte/macrophage colony-stimulating factor (GM-CSF) 13, 19, 20, 76, 86, 88
    mast cells 15, 16
    mouse models 123
    T cells 18
granulomas 38
granulomatosis 42–3
guinea pig inhalation model 109, 112–3
    injection model 110–1, 113–4, 116
    intratracheal test (GPII) 109–10, 112, 116, 136, 138, 139, 143–5
guinea pig models 39, 107–17
    compared with mice 121, 128, 136–7, 138, 139–40, 143–5
    regulations 152, 157, 162, 163
    sensitization 63, 86, 87, 107–17, 135
guinea pig serum albumin (GSA) 111

hair care industry 32, 36
hay fever 8, 10, 74
hazard identification 154–8
Health and Safety Executive (HSE) 61, 68
    regulations 152, 160, 161

hemidesmosomes 20
hexachlorophene 36, 47
hexachloroplatinic acid 114
hexahydrophthalic anhydride 2
hexamethonium 115
hexamethylene diisocyanate (HDI) 2, 34, 43, 124
histamine 8, 12–3, 14, 17, 18, 49, 107
    occupational asthma 39, 42
homocytotropic antibodies 2, 77, 80, 83, 85, 90
house dust mite (HDM) 10–1, 21, 82, 122
hydralazine 47
5-hydroxytryptamine 123
hypercholinergic-induced NSBH 38
hyperplasia 20
hypersensitivity pneumonitis (HP) 29, 31, 38, 42, 44, 73
    diagnosis 49–51
    wood dust 4–5
hypertrophy 20

immediate response (onset reactions) 33
    guinea pig models 107–10, 112–4
    mouse models 123
immunoglobulins
    IgA 44
    IgE 2, 8, 10–3, 22, 43–6, 61–5, 124–8
        diagnosis 51
        guinea pig models 107, 108, 111, 113–6
        mast cells 15–6
        mediated response 8, 33, 37–9, 45–6, 51, 64–5, 107
        mouse models 122–8, 131, 135–45
        regulation 152, 154, 157
        sensitization 43, 74–7, 79–90
        T cells 17–8
    IgG 18, 37, 43, 44
        guinea pig models 110, 112, 113
        mouse models 122, 123
    IgG1
        guinea pig models 108, 112, 113, 115
        mouse models 137–45
    IgG2a 80, 81
        mouse models 123
    IgG4 2, 44, 74, 75, 76, 79
    IgM 18, 37, 43
inflammation 7–22, 135–6
inhalation model 109, 112–3
intercellular adhesion molecule-1 (ICAM-1) 15
interferon-γ (IFN-γ) 14, 17, 18, 38
    mouse models 123, 124, 129, 130
    sensitization 75–83, 86
interleukins (IL) 13–4
    IL-1 14, 15, 20
    IL-2 17, 21, 38, 76, 77, 123
    IL-2R 21
    IL-3 13, 15, 16, 19, 76, 79, 81, 86, 123

Index

interleukins (IL) (*cont.*).
  IL-4 15–8, 22, 38, 75–82, 86–7, 90, 123–4, 129–30
  IL-5 13–9, 22, 38, 42, 76–7, 79, 81–2, 86–7, 123
  IL-6 15, 16, 18, 20, 76, 123
  IL-7 76
  IL-8 13, 15, 16, 20, 76, 88
  IL-9 18
  IL-10 15, 16, 38, 76–81, 123–4, 129–30
  IL-12 78, 82
  IL-13 15–8, 38, 75
interstitial fibrosis 29, 51
interstitial pneumonitis 46, 52
interstitium 42
intranasal mouse model 139, 140–4, 145
intratracheal model 136–40, 144–5
  compared with intranasal 140–5
  see also guinea pig intratracheal test (GPIT)
ipratropium bromide 20
irritant-induced asthma 32–3, 47
isocyanates 1, 11, 62, 63–5, 67, 74, 81, 122
  guinea pig models 112, 114
  regulations 152, 155, 157, 161, 162
isoniazid 47
isotype switching 18

labelling 154–8
β-lactam 47
Langerhans cells 83, 88
late asthmatic response (LAR) 12, 13–4, 15, 40, 42, 85
  diagnosis 49
late onset reactions in guinea pigs 107–9, 112
late respiratory systemic syndrome (LRSS) 37
latency periods 32–3, 47, 61
leucocyte function antigen 1 (LFA-1) 14, 15, 18
leucocytes 9, 13, 14, 15, 37, 41, 85
leukotriene antagonists 22
leukotrienes 8, 13, 14, 19, 85
  LTE_4 18, 42
lipase 144
lung diffusing capacity (DLCO) 51
lymph nodes 80–1, 84, 88
  mouse model 123–5, 129–31, 143
lymphoblasts 37
β-lymphocytes 17, 18, 75, 76, 90
T-lymphocytes 10, 15, 17, 21, 31, 37, 40, 76–83, 85–7, 122, 129
lysozyme 10

macrolide 47
macrophages 12, 20
  alveolar 40–1, 84, 88, 89
major basic protein (MBP) 20, 85
major histocompatibility complex (MHC) 43, 77–8, 82, 84
maleic anhydride 2

maple bark disease 45
mast cells 8, 10, 15–6, 38
  EAR (early asthmatic response) 12–3
  mouse models 123
  sensitization 74, 79, 80, 81, 85, 86
  T cells 16–8
material safety data sheets (MSDS) 49
maximum exposure limit (MEL) 159, 160
mesothelioma 1, 30
metal fever 46
metal industry 29, 32–7
metallic salts 45–6
metalloendopeptidases 20
methacholine 41
  challenge test 49
methenamine 155
methyldopa 36, 47
methyl methacrylate 155
monkeys 15, 137
mouse IgE test 124–8, 131, 143
  regulations 152, 162, 163
mouse models 114, 121–31, 135–45
  regulation 152, 157
mucous membrane irritation syndrome 44
mucus plugging 9, 13
multiple chemical sensitivity (MCS) syndrome 47–8
myofibroblasts 20

National Exposure Survey Data Base (NOES) 31
National Institute for Occupational Safety and Health (NIOSH) 49, 68
natural killer (NK) cells, 78, 79
neurokinin A 39
neuropeptides 20, 39
neutrophils 13, 14–5, 16, 49
nickel 35, 37, 45, 82, 155
nitrogen dioxide 12, 88
non-specific bronchial hyperresponsiveness (NSBH) 32, 33, 38, 39, 40, 46, 47
  diagnosis 49–50

occupational asthma *see* asthma
occupational exposure limits 158–61
occupational exposure standard 159
Occupational Safety and Health Administration (OSHA) 31, 158
opiates 8, 37, 39
organic dust syndrome 44, 49
Organization for Economic Co-operation and Development (OECD) 152, 158
organophosphate insecticides 38
ovalbumin 88, 90
  guinea pig models 108, 109
  mouse models 135, 139
ovomucoid 139
oxazolone 80, 116, 124, 129
ozone 12, 38, 63

P-selectin 14
paraphenylene diamine 36
passive cutaneous anaphylaxis (PCA) 108, 110, 112–4
peak expiratory flow rate (PEFR) 50
penicillamine 47
penicillin 11, 35
persulphates 36, 47
phenylglycine acid chloride 36
phosphodiesterase (PDE IV) inhibitors 21
phthalic anhydride (PA) 2, 34, 38, 39, 65, 80
  guinea pig models 112, 113, 114
piperazine hydrochloride 36
plastics industry 34, 43, 44
platelet activating factor (PAF) 14, 18, 22, 42
platelet derived growth factor (PDGF) 20
platinum and platinum salts 1, 2, 11, 31, 35, 39, 45
  occupational asthma risk 63–4, 65, 69
plicatic acid 2, 37, 38, 42, 63
  wood dust 4–5
pneumoconiosis 1, 30, 49, 154
pneumonitis 38, 45, 46, 52
  see also hypersensitivity pneumonitis
pollen 10, 11, 81, 82, 90, 122
pollution 87–90
polyisocyanates 38, 43–4
polyvinyl chloride (PVC) 11, 47
potroom asthma 33, 46
pranlukast 22
primary prevention 65–7
printing industry 32
procion yellow 112
prostaglandins 12–3, 43
prostanoids 8, 18
protease 10, 116, 144
  B7 proteins 77
psyllium 36
pulmonary haemorrhage and lesions 37
pulmonary infiltrates with eosinophilia (PIE syndrome) 29

quaternary amines 42

radioallergosorbent test (RAST) 43, 44, 45
rat basophil leukaemia (RBL-2H3) cell
  serotonin, release assay 137–40, 142, 143
reactive airways dysfunction syndrome (RADS) 32–3, 44, 47
reactive dyes 2, 47, 112, 114, 116
reagin 8, 45
red cedar wood 32, 35, 38, 44, 63, 67
regulations 151–64
respirators 51
respiratory syncytial virus (RSV) 12
rheumatism 8
rhinitis 39, 45, 61, 107, 155, 163
rhinoconjunctivitis 39
risk factors and management 3, 10, 31–2, 52, 61–9, 87, 89, 91, 158–62

rosin 155
rye grass 10, 11, 136

salbutamol 11, 36
salmeterol 21
scleroderma 43
screening tests 65–7, 152
secondary prevention 67–8
Sentinel Health Notification System for Occupational Risks (SENSOR) 31, 68
sequoiosis suberosis 45
serotonin 79, 137
  RBL release assay 137–40, 142, 143
sick building syndrome (SBS) 30, 48–9
silicosis 43
slow-reacting substances of anaphylaxis (SRS-A) 12
smoking 8, 11–2, 31–2, 45, 48, 63–4, 66
  maternal 11, 89
  regulation 153
  sensitization 89–90
smooth muscle 9, 12–3, 17, 20, 40, 85
sodium cromoglycate 8, 16, 18
spiramycin 36
stem cell factor (SCF) 15, 16, 20
styrene 47
substance P 39
subtilisin 63, 109, 110, 111, 136
sulphonechloramides 36, 46
sulphur dioxide 12, 38
Surveillance of Work Related and Occupational Respiratory Disease (SWORD) 1, 30, 31, 68
  regulation 154, 160

T-cell receptors (TCRs) 10, 17, 77–8, 83
Tc1 cells 82, 89
Tc2 cells 82, 83, 89
Th (helper) cells 76, 77–9, 81, 82
  mouse models 123, 129
  Th0 cells 77
  Th1 cells 17, 38, 76–82, 86, 89, 123, 129
  Th2 cells 17, 18, 38, 76–83, 86, 89, 123, 129
Thp cells 77
T-lymphocytes 10, 15, 17, 21, 31, 37, 38, 41, 76–83, 85–7, 122, 129
  aleolitis 42
  sensitization 74–88, 91, 122
tachykinins 39
tertiary prevention 68
tetrachlorophthalic anhydride (TCPA) 2, 31, 34
  occupational asthma 62, 63, 64, 65, 67
tetracycline 36, 47
textile industry 29, 32, 33
theophylline 20, 21, 22, 52
thermophilic actinomycetes 42
threshold limit values 110, 116, 136, 158
thromboxane 12, 22
thymus 11, 76

Index

toluene 34
toluene diisocyanate (TDI) 2, 37, 39–43, 68
   guinea pig models 111–14
   mouse models 124, 129
trimellitic anhydride (TMA) 2, 34, 37–8, 42, 44, 65, 80
   guinea pig models 112–7
   mouse models 122–7, 129, 130
trimeric hexamethylene diisocyanate (Des-N) 112, 113
tropolones 44
tryptase 12, 17
tuberculosis 17
tumour necrosis factor-α (TNF-α) 88
tumour necrosis factor-β (TNF-β) 14, 15, 19, 20, 38, 76
tungsten carbide 35

urea formaldehyde 36
urticaria 61

vanadium 11, 35
vanadium pentoxide 46
vascular cell adhesion molecule-1 (VCAM-1) 15, 87
vasoactive amines 85
vasodilation 9, 12, 74, 85, 86
verlukast 22
very late antigen 4 (VLA-4) 15, 18, 22, 87
viral infections 12, 21, 153
volatile organic compounds (VOC) 48

wood dusts 1, 44–5
wood industry 1, 35, 44–5
   see also red cedar wood

zafirlukast 22
zileuton 22
zinc 6

WF 150 T755 1997
Toxicology of chemical
respiratory hypersensitivity